The Intelligence behind 'Nature'

Transformation from energy to matter
How consciousness becomes Manifestation

Daniel Perret

Behind all manifestation there is a consciousness,
that is a sentient, intelligent being.

My thanks go to Bob Moore and C.
Thanks to Marie Perret for corrections and editing suggestions.

© 2021 Daniel Perret
Edition: BoD - Books on Demand GmbH
12/14 rond-point des Champs Elysées
75008 Paris, France
Imprimé par Books on Demand GmbH
Norderstedt, Deutschland

Dépôt légal octobre 2021
ISBN 9782322380619

Cover photo Daniel Perret

I dedicate this book to Liv and Theo

A materialist cannot prove that nature spirits
and the invisible do not exist.

To believe that a rose, a tree, a koala bear, or that a human
being come only from a seed is a limited concept,
similar to thinking that a sophisticated car or a computer only
come from a shop. We need to take a closer look.

When we look at manufacturing a car, we easily accept that
there are designers, technical engineers, and workers
producing all the parts. It is usually more difficult to imagine
that behind an energy structure or a tree that there are
numerous invisible beings.

While in many ways of course quite different, we can
compare what I will describe to the organization of a town
with its laws, sophistication, government, administration,
craftsmen and women, beauty, art works, artists, and last but
not least its town spirit.

The more we look into nature the more it is likely to surprise us
by its wisdom, intelligence and complexity.

The Intelligence
behind nature

manifesting
consciousness

Typeface of the Irish Tree Alphabet by
Katie Holten
The free font is available to download from the website: www.treealphabet.ie

Contents

Summary

Part 1: explores the conscious intelligence operating through nature, through every tree, river system, or flower.
Part 2: explains what the work of conscious intelligent beings in nature concretely consists of. Through their work we begin to understand in
Part 3: the miracle of consciousness as it becomes physical manifestation.

It is characteristic of us 'civilized people' to deal with nature from a purely anthropocentric point of view, even when this is called eco-centric in the ecologist movement and the Rights of Nature initiatives. It is presumed that in nature there are no intelligent communication partners that can express their own point of view. This anthropocentric stance is based on a shift in consciousness that occurred in the 17th century when Newtonian/ Galilean science consciously decided to split away and ignore the conscious intelligence of nature. A more sophisticated analytical way of thinking was then introduced that opened the way for the more detailed research of physical nature; at the same time this meant losing the connection with the whole and the meaning. This created our whole unbalance and alienation towards nature and life.

I am exploring whether there are intelligent conscious-nesses in nature that we can communicate with, that have a conscious intelligence beyond the well-researched instinctive intelligence. Recent philosophical discoveries confirm that consciousness is based on feelings and does not need a brain in order to operate. This allows us to understand that consciousness is not limited to human beings. Quantum field theory and energy science are possibly converging in the understanding of how energy layers operate and become transformed into matter, one coming from physics, the other from spiritual healing.

Any communication with possible conscious beings in nature needs to fulfill some basic criteria especially when evaluating whether they can contribute to political or legal decisions. A recent interview with the Loire River Elemental can give an idea of how useful such a cooperation might be. I bring examples of what the work of nature spirits concretely consists of and how organized and complex the operation of nature is, managing the transition from non-physical energy/consciousness to physical manifestation. I will conclude by explaining how energy-science explains the manifestation into the physical.

My way in

Whenever my teacher Bob Moore* began to speak about the four layers of the etheric, I was always fascinated and knew I had to write down every word. He was showing us the door to the mystery of Creation.
*) Bob Moore, spiritual teacher, Psykisk Centre, Denmark, 1928-2008

Then there is the personal connection I have with large solitary trees. I always feel that they are noble beings, each with their own personality, their own beauty. Maybe it is because of their size and how they occupy space, the utter harmony in how they display their branches, the clear edge of the tree canopy, as if someone was trimming them neatly.

I vividly remember a German book about flower fairies with exquisite fine drawings, that our mother used to read as good night stories.

Then, about fifteen years ago my friend Eva Høffding started to channel the Mother Earth Being, the Black Virgin, Ignatius de Loyola, and our teacher Bob Moore, who had recently died. I felt the unusual purity of their intervention as if they were in the room. Dialogues with invisible beings suddenly became a natural reality.

This came after I had read many of the extraordinary 40 Flensburger Hefte books with numerous interviews with nature spirits.[4] There was no doubt, that this was pure, egoless transmission.

So, I was somewhat prepared when the College of spirits of Rocamadour – I call them 'C' for college – first contacted me and sometime later many nature spirits of all kinds and sizes wanted me to send them distant healing. Through their requests I learned a lot about their way of operating, their organization, their problems, and nature.   For more details on 'C' see page 155 in appendix.

The twenty years study of subtle energy, meditation and transpersonal psychology with Bob Moore led me to the necessary grounding in reality. It brought me a deep sense of discernment and responsibility. I felt the need to pass this on and to show that we can be grounded and at the same time in touch with non-physical beings. I became convinced that this co-operation was absolutely necessary to help get the planet out of its present day troubles.

The cooperation with 'C', the collegium of spirits of Rocamadour, brought me a tremendous number of insights. I learned to ask all possible and impossible questions and saw that there was always an answer. I have already shared this in a number of books and will build on this here. I feel a great sense of responsibility when exploring and writing about invisible beings. It is not because something is invisible that we can say and write just anything. On the contrary.

In the following pages I will go into great detail at times, in order to show the depth and precision with which nature and consciousness work.

I am writing because I want things to be clearly understood and I miss not being able to write in my own language, which is the Zurich dialect of Swiss German. I grew up in Zurich. I feel that I cannot reach the beauty and richness I would like to with words. Perhaps my deepest way of expressing is when I improvise on the harp.

I have recorded harp improvisations called 'Birth Rights of Nature' to accompany what I write here and it is available on danielperret.bandcamp.com.

Introduction

Creation, as I see it, happens as a co-operation between beings from the Divine Field or Universal Intelligence and beings from Mother Earth. It is a co-operation between the male father Heaven impulse or 'loving intelligence' and the manifestation potential of the female mother Earth aspect 'intelligent love'.

The Latin roots of the term 'natura' comes from *nascor* (being born, coming from) and the suffix *urus* which means 'an action to come'. *Nature* can thus be translated as 'the force that is meant to generate', referring to an order and system of laws that precede creation of all things and beings. [5]

This wide definition of the word 'nature' includes humans. While it thus is getting close to the meaning of the word 'Creation', I prefer using the term 'Nature' as it feels less abstract and brings us close to our recent concerns about the Rights of Nature, the respect of nature as an intelligence, full of wisdom and beauty.

The Rights of Nature

The extreme disharmony that we observe today with climate, biodiversity, deterioration of our environment and resources, is bringing forth a worldwide realization and actions in what is labelled the movement for the Rights of Nature.

It is my involvement with the Global Alliance for the Rights of Nature (GARN) that made me realize the difficulty we 'Westerners' have to find a genuine and real way to connect with the intelligent beings behind nature, behind a river system, a tree, or a landscape. We readily admire native people who keep that tradition alive and who in fact have largely inspired us and for example have brought into their constitutions and laws the recognition of the rights of some rivers.

However, we seem largely stuck in an anthropocentric approach, even when we label it eco-centric. This attitude echoes the colonialist, paternalizing approach even if it can be well meaning. Such an attitude separates us from beings we do something for, but we believe cannot communicate because of not having the adequate intelligence. We simply don't know any more how to perceive and communicate with the intelligent beings that exist behind visible nature. Yet, every country, every people once had an understanding of the intelligence behind nature and would not have survived otherwise. We might not either, if we don't reconnect to that wisdom. It is at hand, waiting for us.

That misjudgment is partly due to our very left-brain culture that believes conscious intelligence depends on a brain and mental faculties, which is not the case as we shall see.

For historical cultural reasons we have divorced ourselves from the connection to nature's wisdom.

Since the 17th century Newtonian science moved away from recognizing any conscious intelligence in nature.
In the 17th century Isaac Newton laid the foundations for classical mechanics. He also contributed to the progress of optics, astrology, and mathematical analysis. What is less known is that he had dedicated much time to theology and alchemy. John Maynard Keynes, the eminent economist of the 20th century, considered that Newton was the last 'magician' and 'alchemist' of the Western world. [6] Newton was thus familiar with both aspects: the exterior material world and the invisible dimension explored in alchemy. My thesis is that laying the solid bases of physics of the visible physical world was fundamental for our civilization and science. These 'solid foundations' had become an absolute necessity in order for science and the world to move away from the superstition, confusion, fears, and obscurantisms of the Middle Ages. There was great confusion about the role of the invisible, how to perceive it and how to under-stand it in a non-egoistic way.

Although this brought about a fabulous progress in analytical thinking, allowing the material part of nature to be dissected and analyzed in great details, it led us to lose the meaning of how and why all was connected. This very thinking divorced us from the wisdom of the native connection with nature …and often made us destroy their ecosystems and ours.

The Western world has taken at least three hundred years to move away from the confusion regarding invisible phenomena and then to learn how to discern.

It is a fact that research into the invisible worlds, into inner feelings, requires us to be anchored in physical reality, able to deal with the physicality of our life's circumstances. Otherwise, we are faced with the projections of our inner fears, and not able to discern friends from foes.

It is only in the 20th century with the discoveries of the theory of relativity, subatomic particles, and quantum physics that science has become ready to explore the invisible dimensions any further.

The fact that the intelligence of nature and of the invisible beings part of it, were pushed into a corner and mostly survived as a memory in fairy tales, now becomes understandable. The scientific world will only fully recover from being stuck in materialism once we will have integrated the discoveries of quantum physics. A change of paradigms is under way. 2000 years ago, all the cultures in Europe: Ancient Greece, Rome, Celts, Nordic countries had for example gods of thunder, of a mountain, and spirits of the wind. These names or concepts corresponded to the understanding of those times.

Ignoring their existence for centuries, we find ourselves in a position of hardly knowing what the possible function and fields of competence of different kinds of nature spirits are. We believe that nature seems to function pretty much without us, but does it? and does it function well? What benefits would a conscious partnership bring?

Questions Biology avoids exploring

Biology takes certain tasks and phenomena for granted and brushed the following questions under the carpet. Since they don't appear in our biology books, we don't perceive them as questions anymore.

What makes a cell, a plant, a tree absorb certain substances and reject others?
Who or what in nature is analyzing, comparing alternatives and makes the choices between substances?
What creates the harmonious order and beauty for the growing of cells into stems, twigs, leaves, flowers?
What places and connects the cells?
What secures the moving of liquids in a tree when there are no leaves yet / anymore?
How does beauty, harmony and balance come about in the shape of leaves, trees, or flowers?
Where do the plans come from that tells a tree to grow a leaf in a certain form, its canopy, its bark, its type of wood?
We all appreciate it in nature but take it for granted.

How are changes concretely being installed in a gene?
Simply saying it is evolution, does not explain the
manipulation in genes.
Is there an intelligent consciousness behind nature going
beyond the well documented instinctive biological
intelligence of plants and animals?
Are there invisible intelligent conscious beings operating
in nature? What would the function of such invisible
beings in nature be?
Can we observe them? Communicate with them?
Would nature function without them?
Do they have a wisdom and knowledge that we need in
order to restore ecosystems and biodiversity?

Biologists usually go as far as calling that 'intelligence',
but do shy away from stating that intelligence means
consciousness and implies a being.

Quotations

I am participating in the European Hub of GARN's
research group and in 2021 we have been working on a
book about the Rights of Nature in Europe. I contributed
a chapter on the topic of 'The Intelligence of Nature'.
For the last forty years I have studied and taught about
subtle energy fields, spiritual healing, meditation,
consciousness, and nature spirits and had come to feel
and know that all is one, all is connected. We can learn
to quieten our busy mind and develop a connection to
the spaciousness and stillness that lie within us. We can
feel that it is not a physical body that produces spiritual

energy, but rather the spiritual source that creates our physical body, part of a continuous flow, life after life.

I occasionally browsed through the internet searching for a clear definition of consciousness and I was thrilled to find Federico Faggin's series of articles on the nature of consciousness in the Wall Street International Magazine. He writes with clarity and depth about the nature of consciousness and how consciousness is essentially based on feelings.

My teacher Bob Moore's main message was the same: all is ultimately based on feelings. My first book had the title 'Music – The feeling Way'. Energy and the spiritual dimension beyond thoughts, beyond words is accessible primarily through feelings. Feelings are the key that allow us to experience the depth of consciousness and to appreciate the intelligence at work behind nature's manifestation. Feelings are the bridge that connect the heart to the non-physical dimension, from nature intelligences to the beings of the Divine field.

Federico Faggin has the gift of being able to explain consciousness, perception, feelings and quantum fields in a clear and simple way. Most of what I read about quantum fields seems limited to mathematics, technical models and particles. Faggin grasps the philosophical implications and his reflections made me wonder whether the 'fields' talked about in the science of spiritual healing and in quantum physics were actually the same: quantum fields being subtle energy fields.

Faggin studied physics and had one of the first businesses manufacturing microchips for computers in Silicon Valley. He gradually explored the philosophical dimension of what was the essence of consciousness and thought deeply about why artificial intelligence would never reach the same depth. I was happy to find someone explaining with such competence the links that I already deeply felt and knew to be true. The spark that lit up the whole picture for me was that he came to the same conclusion we can find in the Christ letters:

All manifestation comes from consciousness.[1]

Over the centuries many have channeled, interpreted and often misinterpreted what they thought were Christ's words as a teacher. Texts like that often bring up skepticism, more so than the words of an Italian physicist and computer specialist. Yet, both texts emanate their own quality and are convincingly concordant.

We are living in a universe of meaning and precision as shown in the laws governing energy and consciousness: the law of attraction, of cause and effect, of love, and the law of truth. When we work on making our personal ego filters more transparent, the original picture appears. In the following chapter I will therefore quote Faggin on several occasions as well as the Christ letters.

1st Part:

The Philosophical-scientific approach to a conscious intelligence in nature

Energy and consciousness

In order to grasp how physical manifestation comes about we need to have an understanding of the nature of energy and consciousness. The word 'energy' is in itself a loose term which includes every phenomenon that is not physical matter. Yet, even solid matter is ultimately only composed of energy. When we exclude well known energies like electrical energy, often the term 'subtle energy' is used.

If we want to understand subtle energy, we need to bear in mind that: Behind all manifestation there is a consciousness, that is a sentient, intelligent being. [1,2]

"All that you see, all that you touch, hear, feel, know is consciousness/awareness made visible." Letter 5 / p.16 [1]

"There is nothing in the universe that is not consciousness made visible." Letter 5 / p. 17

All energy is therefore a manifestation of consciousness. We want to explore who these sentient, intelligent beings are and why consciousness is always linked to a being. David Chalmers suggests that consciousness should be accepted as another basic universal constant like space or time. [21]

Sentient beings

'Sentient' is defined as being 'responsive or conscious of sense impressions' (Merriam-Webster). Sentient beings are able to experience feelings. The term 'sentience' comes from the Latin term 'sentientem' meaning a feeling. Sentient beings can be physical or non-physical. In a strictly anthropocentric view, whether a being is *considered* to have a feeling, a sentient experience, depends on the human awareness (individual, group, state, law) that defines how that being is perceived.

A sentient being can feel, perceive and sense things. Sentient beings have an awareness of surroundings, sensations, and an ability to show responsiveness. Having senses makes something sentient, or able to smell, communicate, touch, see, or hear. Sentient beings generally have free will and an awareness of themselves. The extent to which they can feel happiness, sadness, pain, joy and fear differs from species to species.

(see page 71, Gnomes, Undines, Sylphs)

There are sentient and non-sentient beings.

What is a being

A being can be defined as created by a will and having an individual existence, material or immaterial, some kind of consciousness, memory, a purpose, and a will – even when sometimes strictly limited - to pursue that purpose.

There are non-sentient rudimentary forms of beings corresponding to that definition, like machine beings (p. 125), sound beings, or thought beings. Each sound produced on a musical instrument and each thought are creating an energy movement or energy structure which becomes a being with sometimes a limited life span. Felt perceptions and feelings (category 1, 3, 4, page 39) don't create such types of 'artificial' beings as they are part of a natural immediate circulation connecting us with nature and All-there-is. The perceptions of a colour, a sound, a taste, a smell, etc. are feelings, or felt perceptions as Faggin puts it. They don't need to create and to live on as a being, possibly because they have no purpose to fulfil other than being part of a circulation in the present. It is only when we produce a thought linked to a felt perception, that a thought being is created.

A thought is created with a purpose: to be expressed, memorized, understood, trigger a reaction, a circulation. Once this circulation is complete, their purpose is met, heard, understood, let go of, it dissolves. The same is true of consciously created musical notes. Their being dissolves after a while when that note is heard and thus fulfils its purpose. Same with machine beings, etc. Once they have fulfilled their purpose and are for instance physically destroyed, the being dissolves.

All beings have the capacity to accumulate **memory**, even if this happens, with some, rather like an automatic unconscious storage of what they do and what happens 'around them' during their existence. They nevertheless keep to a minimum, for want of expression, a means of displaying or sharing the content of their memory.

This leads me to state that a communication based on feelings is a direct, immediate circulation. A 'green thumb' gardener may communicate intuitively with Devas to find the right spot to plant a rose bush or young tree. In the case communicated by the Loire Valley Elemental, a more elaborate dialogue was needed, involving words and thoughts (page 56). Yet, the feeling contact has to be there first in order to allow this type of dialogue to happen.

The unfoldment of Creation, both on a macroscopic and individual level, is built on two impulses: The impulse of activity and the impulse of attraction – repulsion. All impulses are directed by a purpose, in the case of Creation it is a Divine purpose. see also letter 5 / p. 24 and 25

Spirit beings

Spirit beings are non-corporeal sentient entities that possess a conscious connection to the spiritual dimension. This is called 'spirit'. There are far more spirit beings without a physical body than with one. There are a great variety of spirit beings, whereof nature spirits and the angelic beings of the Divine Field are only two kinds.

see also list at the end 134

We can loosely define **nature spirits** as beings created by the 'Earth Goddess' and the beings of the Divine Field as created by 'Father Heaven' or 'universal intelligence'.

While all spirit beings do have **free will** up to a point and are sentient beings, some beings neither have a free will nor a full range of feelings. The question of 'free will' could be developed in more detail: for example, angels do not have a free will, strictly speaking. Their activity is directed by a divine impulse. Yet, it is generally fundamental to an individual sentient being to be able to take decisions about accepting and rejecting. This faculty makes sentient beings individuals, able to pursue their purpose, creatively contributing to evolution and staying alive.

This brief exploration of beings helps to become aware of the importance of a feeling contact and that nature spirits and beings of the Divine Field are something special within the multitude of possible beings.

Energy fields, quantum fields
Quantum physics, and especially quantum field theory, may give us an idea of the deeper mechanisms at work. Federico Faggin speaks of 17 known quantum fields. [2] Each field has its own rules, laws, and characteristics. Each quantum field seems to be a totally different universe from other quantum fields.

That is also how I understand energy fields around humans, plants, and animals. I know little about quantum fields, but the perspective of physics and energy science moving towards each other is very interesting. I have studied energy fields for over forty years. I can observe 17 energy fields around the human body, going from the four layers of the etheric close to the physical body, up to the planetary and cosmic layers. Our energy fields connect us with everything, with nature, with all that happens on planet Earth, with the universe and with our individual timeless soul.see drawing page 128

Each human energy field functions according to its own laws and characteristics and is like a world of its own, which is what makes me compare them to quantum fields. Our mental field for instance functions at a much higher vibration than the astral/emotional field and works essentially with abstract thoughts, words, concepts, and images. Whereas the astral/emotional field operates with attraction/repulsion, painful emotions (for the lower astral) and more refined feelings like compassion, joy, serenity (in the upper astral). They have two completely different languages and types of information even if emotions infiltrate and can trigger 'emotional' thoughts. They manage to do so because they are based on beliefs, which are thought patterns.

The same seems to be the case for the different layers of the etheric. Our etheric energy field is a wonder of Creation which is where the transition from energy to physical manifestation happens. It is a system of four main interconnected layers or fields bringing about physical manifestation. Yet, the tools each etheric field uses are completely different from the other field's characteristics. I go into more details in the third section of this book.

Nature of consciousness

Federico Faggin published a remarkable series of articles on the nature of consciousness in the Wall Street International Magazine between October 2020 and April 2021. He states that in his observation thoughts are preceded by a brief non-verbal perception, a kind of feeling or 'quale' of a thought presence. A quale (the plural is qualia) being 'what something feels like'. This makes him say that consciousness is based on feelings and not on a cerebral-mental activity. This is a revolutionary statement and completely changes the way that we, for instance, understand nature.

"The capacity to have feelings and understand their meaning is the essential property that 'explains' how we know. This is the crucial capacity of consciousness."
Wallstreet international magazine, 11 OCTOBER 2020, FEDERICO FAGGIN, 'The nature of consciousness' [2]

Faggin: "I think that consciousness is a fundamental property of the quantum fields not yet acknowledged by physics. What I mean is that each quantum field is a conscious entity, with an inner reality and free will, communicating with the other conscious quantum fields." 'Entity' is another word for 'being', both normally used to describe a non-physical, invisible being.

"Consciousness does not exist in the particles, atoms, or molecules, but in the (quantum) fields and the fields of fields of which particles, atoms, and molecules are states. Thus, the fields are conscious, not the states."

This means that physical units like 'a tree', 'a flower' or 'a river' are not conscious as such, but that the conscious being – the fields of consciousness - behind them are. Conscious beings are essentially individual conscious fields of energy that can have a physical body or not.

" …the existence of our conscious inner reality is impossible to explain with the current 'matter-first' theories. Consciousness cannot arise from matter, devoid of it, any more than electricity and magnetism could emerge from elementary particles devoid of electrical charge and magnetic spin. The current 'scientific' explanation that consciousness arises from complex organizations of unconscious matter is completely inadequate because complexity has nothing to do with consciousness. The only reasonable possibility to make progress is to postulate that consciousness is an irreducible property of nature."
Wall Street International Magazine WIM, 11 DECEMBER 2020, FAGGIN: Consciousness is fundamental [2]

On conscious intelligence in nature
Beyond its well-researched instinctive intelligence, nature has a conscious intelligence. Biologists have described countless examples of species with an extraordinary intelligence using strategies and complex characteristics that allow them to survive in sometimes hostile

environments. One could describe these characteristics as being automatic, instinctive, biological mechanisms resulting from a long process of adaptation and evolution of that species. I will not comment further on this instinctive aspect of the intelligence of nature, essentially based on the memory of learned responses, but will concentrate on how conscious intelligence differs from this instinctive intelligence, what the conscious intelligence of nature is based on and what it consists of.

The respect we feel towards nature reaches beyond its physical manifestation. We know very well that there is a fundamental difference between a train station, a chair, a monument, and the manifestations of nature. The latter have a dimension that move us through their beauty, atmosphere, and presence. That is why being in nature gives us so much.

Yet, there is another dimension to this: A train station or chair don't have feelings, nor an awareness of themselves, nor free will, all characteristics of a conscious intelligence. Beyond the feeling level, nature has resilience, adaptability, and patience. We need to understand what this aspect of the intelligence of nature consists of and makes nature behave as it does. "Consciousness is a fundamental property of nature" [2] I'll be looking into this in greater detail.

When we argue that all manifestation originates with a conscious sentient being, we need to make one additional distinction. The famous white marble statue of the Venus of Milo was obviously created by a sculptor, an intelligent conscious being who is no longer alive. A tree or a river system were also created by a conscious sentient being in the past but they still have a present-day spirit or consciousness with them, shaping them, keeping them alive.

What importance do intelligent consciousnesses in nature have for the Rights of Nature?

As long as we are convinced that nature consists only of objects with, at best, an instinctive intelligence, with beings devoid of conscious intelligence and free will, we will continue to treat it as such. Communicating and cooperating with nature spirits opens the possibility of drawing on their wisdom. They can teach us how nature is operating and what impact human activity and other changes actually have upon them. There are many dimensions that we, as human beings, do not have access to and which these beings do. Nature is *their* domain of expertise. A cooperation would strengthen the arguments and the effects of any political and legal decision.

Despite the considerable efforts science has made, the functioning of and balance in nature remain of an immense complexity that is hardly understood. In Part Two I will describe some of the work of nature spirits so that we may get some understanding of their fields of competence. Each of us have our own fields of expertise.

How does an intelligent consciousness in nature function? How can we connect to and communicate with it? What political implementation would that have, for the Rights of Nature movement and for ecological initiatives about such issues as climate, biodiversity, or pollution?

As Federico Faggin writes: "Physics currently describes a universe that is holistic and dynamic but deals only with exteriority: measurable events occurring in space and time. To explain the existence of consciousness and free-will, we need to address interiority as well."
WIM 11 FEBRUARY 2021, FAGGIN, One and the consciousness units [2]

"Nature is not matter only. She is also spirit." C.G. Jung,
paragraph 229, Alchemical Studies

Max Planck, Nobel Prize in Physics on Quantum Theory: "There is no matter, only a tissue of energies, that an intelligent Spirit has given form. This Spirit is the Origin of all matter." [6]

"But since there cannot be spirit in itself either, rather every spirit belongs to a being, we must necessarily accept spirit beings."

Max Planck then continues by saying that also spirit beings don't come from nowhere, therefore we have to accept that there is a God, a Creator, a Universal Intelligence.

"I regard consciousness as fundamental. I regard matter as derivative from consciousness. We cannot get behind consciousness. Everything that we talk about, everything that we regard as existing, postulates consciousness." [6]

If we want to understand the process how spirit or consciousness become matter and how life in a tree or plant is kept alive, we need to have to explore what exactly is meant by 'God'. I am looking into this concept of 'God' or Creator on page 60-64 describing in seven principles how Creation comes about and originates in the 'Universal Intelligence'. Calling the Creator impulse 'Universal Intelligence' is more neutral than the term 'God' that we tend to personify and which is obscuring its deeper reality. In that we are coming closer to the Buddhist concept of non-personified creative forces.

Max Planck's and Jung's terms 'Spirit' need to be looked at more closely. They obviously use the words 'spirit', 'spirit being' and 'consciousness' synonymously.
I am not certain how they define spirit.

Consciousness does not need a brain
Consciousness is essentially based on feelings. Consciousness is in the first place a consciousness of oneself as a separate individual, different from others and from the environment. This creates a consciousness

with the faculty of perception, the ability to perceive an event, an intruder, an exterior substance, etc. All this defines what a 'conscious being' or 'spirit being' is (in Planck's words) and why consciousness can be defined as always being linked to a being. This starts with any organism and continues down to a cellular level. At any level the organism can decide which substance is beneficial and assimilate it or on the contrary reject and eliminate the substance.

"Consciousness is in fact inextricably linked to a self. It is one of the core properties that characterizes a self: it is its capacity to perceive and know through feelings, through a sentient experience. But there is more, because consciousness can also turn toward the self, allowing it to know itself in addition to perceiving and knowing the outer world. The other fundamental property of a self is the capacity to act with free will."

WIM 11 OCTOBER 2020, FAGGIN, The nature of consciousness [2]

My spirit contacts 'C' say they would like to differentiate Faggin's statement. (This in fact underlines how precise and thoughtful 'C' is). The very small elementals*, Gnomes, Undines and Sylphs, only have a partial sentient experience. They can make a feeling distinction about what is beneficial for fulfilling their task and what is not. They do not possess the range of feeling pain or higher astral feelings such as joy. Salamander (the fire elementals) do have a fuller range of sentient experiences. Salamander cannot feel pain and joy either, but have a sense of spiritual feelings, what spiritually is right, which the other three small elementals do not have. This faculty of the fire elementals is due to the particularity of the element of fire, which is linked to light, truth and the connection to spiritual impulses.   *) see pages 73

"Perception", according to Faggin, "is the conversion of a signal into a feeling, thus a felt experience. The conversion from signals to qualia is called perception. The conversion from qualia to meaning is called comprehension." A quale (the plural is qualia) means "what something feels like."

WIM 11 NOVEMBER 2020, FAGGIN, Qualia, perception, and comprehension [2]

This process includes the faculty of discernment, that makes it possible to compare two events, two objects, an intruder, foreign substance, etc. Then there is the faculty of deciding and choosing between what the being considers to be beneficial and wants to assimilate and what it wants to reject, and eliminate from its system. This implies the quality of free will and a sense of finality: wanting to stay alive, pursue their existence as an individual and to express itself through choices, preferences, and actions.

There are different degrees and nuances of perception, of feelings and decisions. All these faculties can occur on the level of feelings alone, basically no mental-cerebral activity is required. The latter is reserved, speaking on a strictly physical level of manifestation, to human beings. We do not know enough about the mental capacities of non-physical entities or beings. Consciousness consists essentially of feelings.

Feelings operate on different levels

1. Category: body sensations
2. Category: instinctive basic painful emotions
3. Category: refined feelings (like joy, compassion)
4. Category: subtle, spiritual, or higher feelings

Being so used to having a brain and an intellectual activity based on thoughts, it may take some time to let this sink in. Our thought activity can at times be so invading, like a constant inner radio speaker, that it takes practice to find moments, where this thought activity sinks into the background and allows us to become aware of the feeling dimension of consciousness. Once we have this experience, we can easily see how animals and any conscious being can operate all the cognitive perceptions on a feeling level.

The difficulty of recognizing that consciousness is based on feelings comes from our cultural bias or habit of attaching so much importance to thoughts and intellect. Our education and school systems are based on intellectual learning rather than feeling. I insist on this point, as this is essential in order to follow the argumentation I propose.

Consciousness, inherently part of a sentient feeling being, has the possibility of being active. This being can create, structure, organize and develop, even if this happens on levels with different degrees of sophistication. Such a being is not alone. In what Suzanne Simard observes in nature, it is part of a complex organization, embedded within a hierarchy of beings. [7] Her recent research explores the ways in which 'mother trees' recognize and support their kin by passing on genes and substances to their offspring. This makes me think of Federico Faggin speaking of a hierarchy of quantic fields forming the 'big One'. WIM 11.1.2021 FAGGIN, The nature of the self [2]

Developing on from Suzanne Simard's findings (Finding the Mother Tree), it is possible to imagine that these beings can be assisted by an adviser or elder, and by other beings that pass on impulses and directions, helping them to orient their actions and decisions. They can themselves also have the role of adviser/elder to 'younger' or less experienced beings, thus forming a hierarchical organization based on competence and experience.

A whale or a bear have more sophisticated levels of perception, feelings and decisions compared for instance to those of a simple earth worm. The works of Peter Wohlleben [8], Prof. Dr. Suzanne Simard [7], Stefano Mancuso and Alessandra Viola [9] give us an insight into the intelligence of trees and plants. Any vegetable can make the choice to assimilate substances or to reject or eliminate them. They can take more complex decisions as well. Although these researchers do not directly mention being aware of spirit beings behind a tree when describing how trees communicate and warn each other through scents and a root/fungus 'wood wide web', or 'mother trees' fostering their offsprings by sending them well dosed amounts of nutriments, they describe functions of a conscious, sentient being.

The essential thing to understand is that the physical manifestation of a tree would immediately die if its spirit part was withdrawing. Whether we identify the conscious being as being the tree trunk, the branches and leaves or rather as a separate, invisible being, depends on our perception and knowledge about the energy fields of trees or human beings.

What is conscious intelligence

Wikipedia [10] : "*Intelligence* has been defined in many ways as: the capacity for logic, understanding, self-awareness, learning, emotional knowledge, reasoning, planning, creativity, critical thinking, and problem-solving. More generally, it can be described as the ability to perceive or infer information, and to retain it as knowledge to be applied towards adaptive behaviors within an environment or context."

When we follow some parts of the above Wikipedia definition, we come to basically the same conclusion of what can be carried out by any conscious being, visible or invisible. Basically, no brain is needed, just the capacity to differentiate. As Faggin puts it: "…how do you know you had a thought?" you may recognize that you felt something vaguely cross your mental 'screen', leaving a faint 'image', a quale, with the essential meaning of that thought. Most people do not recognize the barely conscious quale-image prior to its swift and automatic translation into mental words (symbols), believing their thought came directly in verbal form. We have usually learned to ignore the qualia that are the essence of a thought."

WIM, FAGGIN, 11 OCTOBER 2020, The nature of consciousness [2]

When Faggin uses the term 'mental screen', we must be aware that 'mental' is more than just intellectual thoughts, but does include the much larger consciousness of the mind. Indeed, the 'upper mental' (p. 136) includes impulses, intuitions, perceptions of symbols, etc. all perceptions based on higher feelings or higher type of qualia. Again: consciousness does not need words, nor a brain.

Reason is the capacity of consciously making sense of things, applying logic – non-contradictory deductions - and adapting or justifying practices, institutions, and beliefs based on new or existing information. *Logic*, ('possessed of reason, intellectual, dialectical, argumentative') is the systematic study of valid rules of inference, *Critical thinking* is the analysis of facts to form a judgment. [10]

All of this can be done by the intelligent invisible beings behind nature. Most of these activities of discernment and analysis do not need mental-cerebral activity. They function just as well using feelings as a means of differentiation. Remembering that consciousness is essentially based on feelings and not on mental activity.

We can distinguish the cognitive, perceiving side of consciousness and the level of intelligence the being applies to discern, evaluate, decide and act upon its choices.

Faggin accurately explores the rich spectrum of levels of feelings and how they allow any conscious being to know their needs, boundaries, and connections with the ecosystem. (see page 39)

WIM, FAGGIN, 11 OCTOBER 2020, *The nature of consciousness* [2]

Also, in the following definitions of Intelligence we can see that not thoughts but consciousness is needed:

"The ability to deal with cognitive complexity"

Linda Gottfredson (1998). *"The General Intelligence Factor" (PDF). Scientific American Presents.* [11]

"The aggregate or global capacity of the individual to act purposefully, to think rationally, and to deal effectively with his environment"

David Wechsler, (1944). *The measurement of adult intelligence. Baltimore: Williams & Wilkins.* [12]

'to *think* rationally' can be replaced by 'process information rationally'.
Intelligence is "…the resultant of the process of acquiring, storing in memory, retrieving, combining, comparing, and using in new contexts information and conceptual skills".

Humphreys, L. G. (1979). *"The construct of general intelligence".* [13]

"The current 'scientific' explanation that consciousness arises from complex organizations of unconscious matter is completely inadequate because complexity has nothing to do with consciousness. The only reasonable possibility to make progress is to postulate that consciousness is an irreducible property of nature."   WIM 11 Dec. 2020 [2]

"… consciousness and free will are properties of certain coherent quantum organizations existing entirely in the quantum world, which I refer to as 'conscious entities'. To interact within the classical world, these entities need a quantum-classical system, a 'body', that can interface with the classical information of the objective classical world on one end and with the quantum information of the quantum entity on the other end. As far as I know, the only quantum-classical systems capable of performing this remarkable feat are living organisms — known to support consciousness already."   WIM 11 April 2021 [2]

The bridge between quantum field theory and energy science could be imagined by replacing in the above quote the word 'quantum' by 'energy'.

Conscious intelligence in nature is superior to 'artificial intelligence'

A brief comparison with 'artificial intelligence' allows us to grasp more clearly what conscious intelligence in nature consists of.

Nature's intelligence is far away from mere computer 'intelligence'. "There is no evidence that electrical patterns in computer memory, or electrical signals traveling along electrical wires, no matter how complex they may be, can produce qualia. In a robot, these electrical patterns may produce reasonable and appropriate automatic responses, and the imitation of a conscious behavior may be so skilled that it deceives us into believing that the robot is conscious. Yet **robots have no consciousness**. They simply do what they are programmed to do, or what they have learned through their artificial neural networks created by conscious human designers with the explicit intention to imitate human behavior. In our physical world, consciousness only exists within living organisms.

Robots have no sensations, **no feelings**, no self-knowing, no free will, and no meaning because these qualities cannot come from statistical matter devoid of consciousness. We perceive and understand only because we feel, and our consciousness is the strongest evidence that we are more than machines. Qualia belong to a different category of phenomena than the physical events we can measure."

WIM 11 NOVEMBER 2020, FAGGIN, *Qualia, perception, and comprehension* [2]

This is a crucial distinction in a time where many seem to be fascinated by AI, not realizing that it leads into a lifeless, feelingless future where robots are created with neither feelings nor empathy.

We can now compare our findings with the above definitions of intelligence found in encyclopedias. No matter how much we go into the details of the definitions presented, we will easily see that these definitions perfectly fit with our definition of a conscious sentient being with or without a physical manifestation.

Communication with nature spirits

The stumbling block for many people is to know how to communicate with the invisible conscious beings behind visible nature. Most people may readily believe there is an intelligence that created nature. Yet, they may limit themselves to thinking that this Creation process took place sometime in the past. They cannot imagine that this intelligence is still present behind all manifestation, being it a tree, a flower, a river system. I will now continue to explore, step by step, what a conscious intelligent being is capable of.

The characteristics that allow communication between humans and conscious beings in nature

A conscious intelligence able to perceive, assimilate or reject, is thus able to perceive signals from other conscious beings such as humans. It is therefore not surprising that such a being can participate in a communication. The feeling contact is essential. Communication cannot be looked for only on a mental level. I remind you now of what I looked into on page 28. Depending on the being and the interviewer, different levels of communication are possible. This can for instance, extend from a simple YES and NO response signifying either acceptance or rejection, up to a complete dialogue where whole sentences are received, as is the case in the Flensburger Hefte Interviews. [4]

This type of communication with an invisible partner raises a number of questions. A rigorous and particularly adapted methodology is needed. Invisibility is a particular condition but not an insurmountable problem. I have been researching in this field for over 40 years. The many communications I have had with nature spirits have consistently shown the following qualities that could contribute to a basis in useful communication: coherence, precision, consistence over time, good common sense, free from ego, respect of our free will, clear, practical, brief, useful to the ecosystem and bringing innovative and often unexpected insights.

See also first page on my website.

The degree of complexity and responsibility involved in a communication depends on the size of the ecosystem and on the political processes involved. When it involves moving a plant or bush in a private garden from a place where it was unhappy to a place where it can thrive, it may only require a person 'with a green thumb'. They will intuitively feel the right place and 'only' the chief nature spirit of the bush or plant is involved in this.

However, when it comes to establishing legal rights for a whole river system or to know how to bring nature back into balance in a wetland or in any other larger area that needs to be preserved, we might want to contact a large water elemental or the landscape angel of the area. We have seen that these beings can be expected to perceive what we do and ask. Yet, how can we be sure of who we are dialoging with and of the quality of our communication? 'Who is on the phone, please!'.

See interview with the elemental of the Loire River System page 56

Prerequisites for communication could be the following: sentience, conscious intelligence, perception of the partners questions, a degree of free will and having an opinion or perception of a choice in a given situation, choosing an answer and willingness to communicate.

Methodology and the political process
The answers and the level of dialogue in such a situation would become more complex when a possible political implementation is involved and thus the need for politically convincing arguments. A decisive aspect is that these invisible beings have no power of decision in human processes. They have the role of consultants while we are keeping our free will.

Some specialized people, let us call them 'mediators' know how to carry out this type of communication. These communication techniques can be learned but this may take some time. This is not unusual. Science in all domains uses at times highly skilled instruments and techniques that need to be learned. It may take some time to build up a culture of communication with intelligent beings in nature and finding competent mediators. Accepting that there are such beings in nature is a decisive first step and makes them feel that we care and respect their knowledge and wisdom. In my experience these beings have always been ready for a cooperation with humans. It is their respect of our free will that has led them to be so patient.

There needs to be an understanding of quality criteria, and this can be summed up as having an egoless motivation, that is the absence of any search for personal power and rewards, and this is basically the absence of fear. It is a question of deeper discernment, not dissimilar to the 'medieval scientific dilemma of Newton' mentioned above. We need to know how to overcome the fear of the unknown and of the invisible.

Dealing with an unknown field can generate uncertainty, possibly feelings of vulnerability and various degrees of fear due to a temporary loss of known reference points. Being aware of this allows us to find grounding in building up understanding and knowledge through systematically exploring the new field. The best antidote to fear remains a combination of understanding and empathy. Understanding needs to include a higher perspective, getting an overview and seeing the larger context. This includes contacting the intelligence of the universe or nature. The steps are: good grounding, good common sense, a caring, empathic connection to nature and towards our own instinctive heart wisdom.

We don't actually need to see these nature spirits in order to communicate with them. The presence of their energy body, usually a column of energy, can clearly be felt with subtle perception and through using certain radiesthesia tools. What we might want to know is where the nature spirits are located, what their field of competence is, what problems and insights they want to share and whether they feel understood by us.

Traditions of communication with non-physical beings

Native people, first nations
Native peoples throughout the world, who have maintained their ancestral links to their land, know that beyond all physical material appearances there is another dimension, a conscious being. Any conscious being is per definition an immaterial system, and can very well exist without a physical manifestation. As we have seen above, the process described as consciousness of self, with a sense of finality, perception, discernment, choice, possibility of action is not limited to physical manifestations. Insects, plants, animals, and human beings, are examples of life manifestations. However physical manifestation is not a necessity for the existence of a conscious being. Conscious beings can remain invisible. This is generally the case for all elemental beings, from the smallest to the largest being, for instance those in charge of the fluvial system of the Loire, the Rhine, the Danube, Rhone, or the Dordogne River. These river systems cover huge areas including all affluents and lakes.

The pitfall of neo-shamanic approach and copying first nations
Copying native people who have kept this knowledge alive over centuries, is not a solution. We can respect and learn from them, but not copy them, as brilliantly described in the article by Dr. Jacques Mabit: *Amazonian shamanism and the Western world: between encouragement and warning* [19].

We don't possess the traditions and ways of understanding of any native people. These are always linked to their very long history and geographic area and cannot be transposed elsewhere. We, here in Europe, have to find a way of cooperating that fits into our present-day culture and issues. Very likely, this includes a more analytical and detailed approach ('loving intelligence', see p. 58), combined with a holistic understanding of an eco-system ('intelligent love').

"We lived off of the bounty of the forest, the bounty of the rivers, and the bounty of the oceans. And it was a holistic lifestyle, in the sense that all things had a living spirit. And we were given a way of life, to live and balance in harmony with those spirits."

Zunigha, Delaware Tribe [23]

"…we must remember our responsibilities as humans to protect the more-than-human communities which sustain our 'narrow conditions of existence'. Without nature, we cannot and will not continue to exist. In establishing the rights of nature for our more-than-human relatives, we engage in reciprocal acts of friendship that are needed to protect the networks of non-humans that support all human life." Chelsea Fairbank and Barbara With [24]

Here is a quote from natives of Canada:
"Our aboriginal people view themselves as one with nature. They don't even have a word for "the environment," because they are one. And they view trees and plants and animals, the natural world, as people equal to themselves. So, there are the Tree People, the Plant People; and they had Mother Trees and Grandfather Trees, and the Strawberry Sister and the Cedar Sister. And they treated them—their environment—with respect, with reverence." [7] Prof. Dr. Suzanne Simard interview in emergence magazine, May 3rd 2021

The use of the term (Tree-)'People', etc. is interesting as it implies the presence of consciousness and the potential of participating in communications.

The following are some examples of the consequences for the political decision processes regarding nature. One could list many more valuable examples like the Saami traditions in northern Scandinavia, and native people of the Amazonas region. When searching for Saami views I read how often they had been disrespected and misinterpreted by official governments and so they became reluctant to share any of their traditional views on nature spirits. I have not found any worthwhile source on the subject yet.

Ireland
In 1999 Eddie Lenihan became well known in Ireland when he persuaded the County Clare Council and the national road authority to move a newly designed road away from a fairy bush near Ennis. The fairies or little elves living in such a bush have an important function in the ecosystem. The inhabitants of County Clare had obviously kept this understanding alive. [25]

Iceland
In the French ethnological magazine 'Ethnologie Française, Vol. 33, of 2003/2004 Vanessa Doutreleau writes [26]: "A major daily newspaper in Iceland describes a problem that had to be resolved concerning the building of a hospital: '...the ground seemed to already be 'inhabited' (by fairy people) (*Dagur*, 13 February

1999). The most famous clairvoyant of the country was asked by the public authorities to come a.s.a.p. to confirm or not the presence of *huldufolk* (a kind of fairy folk). The examples, reported in the first pages of a daily newspaper are nothing exceptional… All bear witness to the more or less widespread interest of the Icelanders in these supernatural beings". "….The *huldufolk* are chtonian beings (mythological earth spirits) found in the whole of the Scandinavian world and belong to the large family of elves (Lecouteux, [27])".

Germany

Over a period of the last 25 years the publisher of the 'Flensburger Hefte', has brought out 40 books of interviews with nature spirits. They have laid the foundation of an enlarged science of biology, bringing precious insights into the little-known tasks that these beings perform in the background. Only two of these books have been translated into English. [4]

Biodynamic agriculture

Researchers in Biodynamic agriculture are used to observing a hierarchy of beings assisting nature. They call them elementals and divide them into belonging to one of the four elements *earth, water, fire, air.* [14] These elementals reach from the very small ones up to a very large one whose territory sometimes covers half of France. There are dozens and often hundreds of these beings at work in a flower or in a tree besides the directing being, that some researchers call a Deva of a tree or of a flower. Having imagined this on the level of a tree allows us to grasp more complex systems like a forest, a fluvial system, a mountain, a national park, a valley, a lake, etc.

Biodynamic agriculture is a form of alternative agriculture very similar to organic farming. It includes concepts drawn from the ideas of Rudolf Steiner (1861–1925). It treats soil fertility, plant growth, and livestock care as ecologically interrelated tasks.

Very small elementals, like gnomes or undines can communicate with human beings up to a certain level, however they do not have an overall view of the plant, tree, or flower. They are highly focused on their specific tasks and fields of competence. We shall see that they cannot communicate with other elementals as these belong to another quantum field altogether (p. 76). If we wish for a more complete understanding of a particular plant or tree we would need to address the 'directing being' of the organism, the Deva. They are found with each flower, plant, or tree, but also as group Devas for a group of flowers or bushes of the same type. I have also mentioned Devas of a whole area, such as a field, or a part of a forest. They seem to adapt by and large to the human organization of a landscape, following for instance limits of properties that manifest in roads, hedges, fences, or changes of vegetation.

We have seen that the landscape angels or regional angels are the most likely communication partners for a large geographical area. The Devas would turn to them for technical advice or problems. (See the list of entities for larger areas on page 134). The hierarchies in nature have the experience and wisdom to bring us the beings we need to contact in each specific situation. There is no competition between these beings as they each have their specific place.

Communicating with nature beings needs to be built up as a practice. Mediators need to be found and trained. The present change of attitude towards nature beings will bring results as they will feel that we respect them. Greeting them is always a useful first step as this shows our change of attitude. The rest will follow surprisingly easily, even if we cannot quite imagine how. They will help us. That is my experience.

I will now present an interview with a large water elemental of the Loire River system in France as I wish to share an example of a communication that I have experienced. I had no mandate from a citizen group to do the interview, so there was no political aim, it concentrated on what that being wanted to share on that particular day. It taught me about the geographical extent of his river system which included the estuary into the sea. The interview also shows that he has constant access to all the details in his vast geographical area.

A dialogue with the Loire River Elemental (June 10th 2021)
A while ago you showed me where you had your permanent anchoring place on the Loire River System, near Bonny-sur-Loire. Is that location still correct? Yes.

Are you a very large water elemental? Yes.

I aim to introduce you to the people working for the Rights of Nature and especially those concerned with the Loire valley, so it would be useful to help us understand what role you could play in this process. Even when I do not have a special mandate to do so at the moment. Is it ok to interview you now? Yes.

How to proceed? Do you have a particular take on the subject? No.

Let me start with my usual opening questions.
Are you well? Yes.
Can I do anything for you at the moment? No.

Would you be interested in a cooperation with people working for the rights of the Loire River System? Yes.

The Loire River Systems includes all affluents from their source to the Loire River? Yes.

Is there a particular section that needs to be looked at today? No.

Is there a particular theme that needs to be looked at today? Yes.

Can we talk about that? Yes.

Does it concern pollution? Yes.
Is it pollution caused by humans? Yes.
By industrial pollution? No
By used waters? No
Tourism? No
Fishing? No
Boats? Yes.

Any other source or type of pollution right now? No
Are we talking about barges transporting goods? No
About individual boats? No
Located near your estuary? Yes
Near Saint Nazaire? Yes
Are we talking about the 'Port de Comberge' at Saint-Michel-Chef-Chef? Yes.
Obviously for you your estuary up to the Pointe de St. Gildas is part of your River system? Yes
The pollution caused by the harbor and the private boats is what you are pointing at? Yes
Is that all you wanted to say today? Yes
Did I understand your answers correctly? Yes.
Thank you. Your indications are unexpected for me and seem completely useful.

As so often, this interview again shows coherence, consistence, unexpected aspects, and is useful, and precise. We had no particular agenda or mandate today. Both would influence the outcome.

Although I receive the answers in a Yes/No form with the help of a radiesthesia device (a small Hartmann-Antenna, see pages 96 and 162), I get the feeling that there always is a non-physical being helping to formulate the questions. The questions are thus almost part of the answer. The interviews always feel much like having an intelligent 'human' on the phone: grounded, direct, practical, non-egoistic. The movements of the device come with the help of the lower life ether: 'Yes' with a swing to the left, 'No' with the antenna remaining stretched straight ahead in its original position. If sometimes I get a hesitant, unfinished Yes swing to the left. Then I know, the question was not precise enough.

Studying nature, and especially through my dialogue with the intelligences in nature, makes me realize how important it is for our times to understand the beauty and complexity of how manifestation in nature happens. My sincere wish is that my work may bring about a way to cooperate more intelligently and respectfully with nature and restore the unbalances that we observe nowadays.

The key to understanding manifestation, the process of becoming physical, is on the one hand to get an idea of what the universal intelligence is based on. There are basic laws and principles at work.

Understanding how the etheric energy fields are operating is the other major key. The etheric is a masterpiece of Creation. In its four levels, or four ethers, we can get a glimpse of how the transformation from consciousness, or from the original impulse, into physical manifestation happens. What is achieved through the four layers of the etheric can feel like 'magic', as it is so complex and astounding to grasp.

Here three quotes from Bob Moore that gave me a lot to think about.

"The etheric, being (for us humans) the subconscious part of mind, is a storehouse of all our activities that we could consider to be good, bad or indifferent. There we have the relationship to emotion, there we have the relationship to the upper area of emotion, the feeling aspect, there we have the mental, thinking activity. There we have the relationship to our individuality, the essence or soul level and there we have also the integration of the spiritual. All can be contained within the etheric."

"The use of these four ethers that we see in the etheric are not just in our etheric, they are also in the etheric around the world. This is why, when we look at the development of any one person, it can never be a singular development, it has got to relate to other people, to the universe."

"The reflector ether is the highest ether, the highest energy contact that you can achieve. In that energy contact you have got all the things that are necessary, you have peace, contentment in yourself, you have your connection with the universe, you have a relationship to light, you have what is drawn from that light into the body through the chemical ether and then you have the balance of life which you will find in the balance with nature and yourself. The reflector ether is reflecting all of that as a combination, but centred really on the peace."

See also page 151

"Quantum mechanics, in its attempt to observe matter in its smallest form, finds that everything is broken down, not into mass, but into energy, where matter can be thought of as a 'slowed down' version of energy." [15]

The Christ Letters from 2000 offer a contemporary explanation of what the Universal Intelligence is. These explanations sound as if they could come out of quantum physics and move us away from an outdated personified Creator figure. The reflexions especially found in letter No 5 give us an idea of the origin of consciousness and what the Creation process consists of. [1]

Some principles of the universal intelligence

1. Principle: **All Manifestation, visible or invisible, originates in a conscious sentient being.**

"Everything in the Universe is consciousness/awareness made visible. '   letter 5 / p. 16/17 [1]

2. Principle: **The Ego made Creation possible.**
The two fundamental impulses of individuality – bonding-rejection created the conditions of individuation, for the creation of individuals, able to exist separate from others.

In plants this individuality is guaranteed by boundaries and the twin impulse of rejection and elimination of substances not belonging to them, and the other pole, the attraction of substances that nourish the plant.
„These impulses of bonding-rejection in the individual personality also become highly creative in that they determine – and make visible – the 'consciousness forms' of 'things desired' and 'things rejected'".   letter 5 / p. 26 [1]

3. Principle: **No Life without Purposefulness**.

"…the membrane of the cell continues to do this work of selection of nutrition and discarding of waste in a billion different circumstances and conditions relating to survival within differing species and differing environments. Is this not evidence of purposefulness displayed within every single action and every single species be it insects, plants, reptiles, birds, animals and human beings?"

letter 5 / p. 9 [1]

"The question is: At what point in Creation did ‚consciousness' creep into living organisms? A consciousness that enables life to select the right nourishment and discern what substances to be eliminated."

letter 5 / p. 12 [1]

How does consciousness get into the DNA? Was consciousness discernible in the living molecules and cells …permitting the intake of selected food and excretion of toxic waste? How is toxic waste recognised? This reflection brings us to the level of atoms, to electrical particles, and to the electromagnetic forces. Where does electromagnetism, which is responsible for the most basic steps in the Creation come from?

Included in this purposefulness is the impulse for more consciousness/awareness that ultimately leads us to go beyond the ego.

4. Principle: **The electromagnetic forces are the basic cornerstones of all Manifestation.**

5. Principle: **Everything is Consciousness / Awareness**.
"The primary comprehensive nature of consciousness is awareness. It is not possible to have consciousness without possessing awareness".   letter 5 / p. 16 [1]

"All that you see, all that you touch, hear, feel, know is consciousness/awareness made visible. There is nothing in the universe that is not consciousness made visible. Consciousness/awareness is infinite and eternal. "

letter 5 / p. 16 [1]

"You are the pinhead expressing Universal consciousness/awareness either in equilibrium when you meditate in stillness of thought – or as active conscious-ness, when you think and feel, plan, and create."

letter 5 / p. 22 [1]

"The ultimate universal dimension of consciousness/awareness can never be fully or truly known by an individualised spirit. It is inaccessible. It is in equilibrium. It is the only source of all power, wisdom, love, intelligence.

The universal dimension of consciousness/awareness in equilibrium is a state of silence & stillness out of which come sound, colour, individualised form, and all visible creativity within the visible universe.

The nature of universal consciousness is intent, inactive and in equilibrium…an infinite, eternal, limitless, boundless state of powerful intent – pristine, pure, beautiful. This intent is to express its nature." letter 5 / S. 17 [1]

6. Principle: **The Universal Purpose is to give individual form to Creation and experience it.**

"... the ‚Father-Mother-Consciousness' is in equilibrium within the Universal Consciousness/Awareness where ‚Father Consciousness is Universal Intelligence/using electricity and Mother consciousness is Universal Love is using magnetism.

The Father consciousness is the **intelligent love** which gives intelligent energy and momentum to the world of complex forms – expressed physically as **electricity**.

Mother purpose is the **loving intelligence** which gives purpose and the impulse for survival to the individualized complex forms – expressed as **magnetism** – bonding & repulsion."
letter 5 / p 19 [1]

"You think with electrical impulses in the brain and you feel with magnetic impulses in your nervous system. They center and bond the electrical impulses into a cohesive whole."
letter 5 / p 24 [1]
'Nervous system' here probably includes also our emotions.

7. Principle: **"The universal dimension is a dimension of unformed impulses. It contains no blueprint of Creation. It is in a state of undivided form."**
letter 5 / p 19 [1]

This leads me to feel that evolution is open ended. Nobody knows where evolution is leading. The purpose does not contain the end result, only the principles of individuation, creative freedom, love, and truth.

"... at the time of the Big Bang 'Father' intelligence & 'Mother' nurturing love were exploded to work independently and also jointly. Their respective 'tools' were Electricity & Magnetism. Out of the explosion of equilibrium came the great intent of self-expression."

letter 5 / S 20 [1]

"The universal awareness of being became the impulse of individualized ‚I'-awareness demanding self-expression. Life and ‚I'-ness are synonymous in the dimension of ‚matter. They became the consciousness of ‚matter'."

letter 5 / p 21 [1]

2nd part

The daily tasks of nature spirits

My personal research into the intelligence of nature

In this section I endeavor to describe how invaluable the work of nature spirits is and why nature simply would not exist or function without their contribution. We mostly take for granted that nature is growing and most of the time working well, but is it? We often cannot imagine that invisible beings are operating behind the visible screen.

My research is based on my observations in nature, of energy structures around plants, animals, and human beings, and on a ten year communication with the college of spirit beings: 'C' (p. 155), as well as my direct communications with nature spirits and beings from the divine field. When passing on information I receive from 'C' I do not, on principle, doubt what is said. These statements come formulated as facts. During the ten years that I have been researching and writing with their help, I always found their contribution consistent, intelligent, based on a sane sense of reality.

During the last five years I have had several hundred exchanges with nature spirits. The communications were mostly brief and based on Yes/No questions related to me through my Hartmann antenna. These contacts brought me some understanding of their tasks and concerns. I learned that they were eager to have contact with human beings and that we could often contribute to solving problems in unexpectedly simple ways. There was nothing extraordinary about these quite matter of fact and pragmatic communications, except that, of course, they were extraordinary.

After an initial phase of about two years where they readily explained exactly where they were located, who they were, and what troubled them, we exchanged less and less of this information. The interaction would limit itself to them asking for help, with no explanation, and then me sending distant healing with the harp or prayer.

I would never have dreamt of being able to help them with a simple prayer, but that is what they mostly asked for. The feedback they gave afterwards was that I had helped them, for me this still remains somewhat mysterious. These communications were based on concrete observations, which were often corroborated by the observations of other people present in the room. The nature spirits' requests came in the form of energy lines converging towards the crystal that I had placed on the floor. These lines would point in the direction of their location. The feedback, that my action had helped them, showed by the disappearance of the energy lines.

I would also ask them if my intervention had been helpful.

These lines would stay there until I had dealt with them. The reason why a crystal was central in this communication may be explained through the use of small crystals in old radio sets. By turning a knob, the small crystal inside the radio set would be directed until it pointed towards the chosen radio station.

Exploring new and unknown fields by definition bring 'extra ordinary' new insights. Western science has been built on paradigms that are not necessarily valid anymore for every scientist.

The building blocks involved in physical manifestation
We need to look at the energy building blocks of physical manifestation and how they are linked together:

- 5 elements: *earth, water, fire, air, space*
- 5 types of elementals: *earth elementals, water el., fire el., air el., elementals of the 5th kind*
- 4 ethers: *chemical ether, light ether, life ether and warmth/reflector ether*
- 4 layers of the etheric: *chemical ether layer, light ether layer, life ether layer, reflector ether layer*

The five elements
In the western, Indian, Tibetan and ancient Greek traditions, and as written by the Swiss Medical Doctor Paracelsus (1493-1541), each of the five elements represent a state of matter in nature. Solid matter is

classified as the "Earth" element. "Water" is everything that is liquid. "Air" is everything that is a gaseous. "Fire" is that part of Nature that transforms one state of matter into another. "Ether/space" can be considered as the energy found in the different layers of energy (or quantum fields) of the human aura (p. 127), the etheric energy field with its four main layers being one of them.

The Chinese system of 5 states of energy is different, but often confusingly also called '5 elements', whereas it is a dynamic system of changing phases of energy. Their 'metal' or 'wood' e.g., describe symbolically a phase of energy, like spring (wood) is an expansive phase or condensation (metal).

I have been teaching musical improvisation for many years using the five elements. [22] They have proved to be a surprisingly accurate way of relating the energy system of a musician with the quality of their musical expression. Some people for example had fluidity in their playing whilst others lacked this quality (water element). Some people needed to bring more 'fire' and 'warmth' into their expression; very few people have spaciousness in their music. In others you could clearly feel generous, elevating melodic movements that felt like birds flying in the 'air'.

The links of each element with a body zone and a psychological characteristic were pointers that helped many people progress by focusing on improving their contact with a weak body zone/element.

The five elements and their qualities have been well known in Indian culture for several thousand years.

Five elements' qualities and body zones

The 5 elements, besides being a physical state, can each be linked in human beings to a body zone and to different qualities that can be felt in music and generally speaking felt in expression.

The element *earth* is linked to our contact to the earth, to everyday reality and to our feet, legs and to the root chakra and to grounding with all its themes forming the bases of our lives. Interestingly enough one can hear/feel all this in someone's expression, being it musical, dance, speech, etc. The earth element is easily heard in the presence or absence of *earthiness* and feelings of security.

The *water* element is linked to the area just above the pelvis and to the hara chakra, an area with hardly any bones except for the spinal column. Once again, one can quite easily hear/feel if someone's expression and being is 'fluid'. Speech would be experienced as naturally and fluidly moving from one thought to the next. The musical quality of rhythm is one expression of this where we would be talking of lively, fluid rhythms rather than mechanical pulses.

The *fire* area of the body is located in the middle of the torso, and includes the solar plexus chakra. The quality in expression would be heard in someone being either held back, too quiet, not 'fiery' and lacking dynamic range of speech or the opposite positive aspects.

The *air* zone of the body is located in the lung area

and includes the heart chakra. Its musical quality would be heard/felt as generous melodies and chords.

Finally, the *space* element is located around the throat and the thyroid chakra. In expression it is heard/felt as presence or absence of silence and space.

We find the four first elements also in our astrological charts as part of the zodiac signs: Taurus, Virgo, and Capricorn being *earth signs*, etc. This understanding is used in biodynamic farming to plant root vegetables during an *earth sign* influence, fruit type vegetables during a *fire sign* influence, and leafy vegetables during a *water sign* influence.

Nature spirits and their tasks

The five types of elementals are linked to the five elements:

- *earth elementals* - gnomes
- *water elementals* - undines
- *fire elementals* - salamander
- *air elementals* - sylphs
- *elementals of the fifth type.*

The elementals are nature spirits, intelligent invisible beings and are usually very small in size. We also find the five types of elementals in larger sizes, each type covering more territory according to its position in nature's hierarchy. Other types of nature spirits can be of different sizes and categories. Very large elementals can have geographical areas of action that can cover areas such as a third of France. The Swiss Medical Doctor Paracelsus described them and Rudolf Steiner also mentioned them at the beginning of the 20th century. The elementals of the fifth type only exist since 1993 and are never very small. Besides elementals, the term nature spirit is also used for a number of other beings with distinct functions.

All elementals, like all sentient being, are sensitive to loving attention, because this brings energy to their upper layer of reflector ether. This upper reflector layer is the interface between human beings, beings of the divine field (mainly angelic forces), and the elementals.

The *fire* elementals function in the life ether, *water* and *earth* elementals in the chemical ether, and the *air* elementals in the outer layer of the reflector ether. The structuring of all cells (shape, alignment in roots, barks, trunks, etc.) is done by the **earth elementals** (gnomes). following the detailed plans from the light ether directives. They have the means of producing particles of matter out of etheric energy. Their building activity can be compared with bees building the alveoli in the bee hives.

The life within and between cells is brought about by the **water elementals** (undines). The water elementals deal with the circulation of fluids. They both nourish and eliminate substances. Water elementals know how to create liquid particles out of etheric energy. They are using this knowledge when helping to move liquids in a plant or tree.

The **fire elementals** (salamander) work in the life ether. In plants the fire elementals are responsible for the maturing of the flowers, seeds, and fruits. They are in charge of the colours, including the chlorophyl process (green), and all warmth & heat processes.

The **air elementals** (sylphs) work in the outer layer of the reflector ether and create, sense, and send out perfumes and scents. They communicate with other plants through the air. The work of Stefano Mancuso and Alessandra Viola [9] documents this.

The *air* element is a carrier of thoughts and light. It is linked to enthusiasm. As wind it spreads pollen. Wide movements of thoughts are carried by the air streams over landscapes. "Sylphs are messengers of the light on a small scale." "..air pollution…is at the same time a pollution of the light. Wisdom corresponds spiritually to light. And when the air is polluted through human deeds …then what the air brings, the wisdom of heaven in the light, reaches the earth in a polluted form." p. 71, [4a]

The four elementals are part of the **magnetic forces,** or the female 'intelligent love' aspect of Creation (p. 63).

While the elementals can be understood as the 'workers' operating within one of the five elements, another type of nature spirits provide the link between these workers and the 'angelic hierarchy or Divine Field'. I call this category '**Devas**'*. Another kind of beings have the function of mediators between different types of nature spirits (I call them the 'large **Elves**'*). They come into action when changes in energy happen and a new balance between the types of beings has to be negotiated.

Another type of nature spirits works on creating atmospheres in places - I call them '**small Elves or fairies**'. Those who help establishing connections between humans and nature spirits I call '**Dagdas**'*. Those who operate between the 'worlds' (or quantum fields) I call '**beings of the magic realm**', etc.
*) other authors may use other names but their function remains the same

Numerous beings from the Divine Field or angelic hierarchies operate within nature but are not nature spirits. I am referring here for instance to angels, who are the immediate link between Devas and the higher beings of the Divine Field: Landscape angels, regional angels, nation angels, etc. (list at the back of the book. and [16])

In recent years a new kind of nature spirit appeared with the specific task of furthering the cooperation between human beings and nature spirits. They are the **elementals of the 5th kind**, linked to the element of *space*. They move around freely. [16]

		The Elementals			
Element	**Earth**	**Water**	**Fire**	**Air**	**Space**
The smallest Elementals	**Gnomes** Roots, bark, Earth, Tunnels, Health of soils Mining Drillings for Oil etc.	**Undines** Liquids, Leaves Stems, Water tap Water ducts rivers and water systems of all kind and sizes	**Salamander** Reifungsprozesse Frucht, Samen Flammen Wärmeprozesse Heizungen Elektrische Leitungen, Computer, etc.	**Sylphs** Scents, vapours, fogs, clouds, perfumes Movements of thoughts	**Elementals of the 5th kind** exist only from the next size level onwards
Some tasks of the small up to very large Elementals	danger of Landslide, Deforestation, tectonic movements, Earthquakes	temperature of the oceans Water Pollution, Flooding, glaciers the large currents in the oceans River systems	Electricity: Production, distribution, consumption Fire, combustion motors, Fire weapons, explosives, nuclear energy, light pollution at night time	Disappearing of insects and birds, air pollution, Bees, Climate change, Wide air streams across countries and continents	Cooperation between humans and Nature spirits. Problems of industrial farming etc.

Here a map of the largest fire elementals and their territory of action in the British Isles and Ireland.

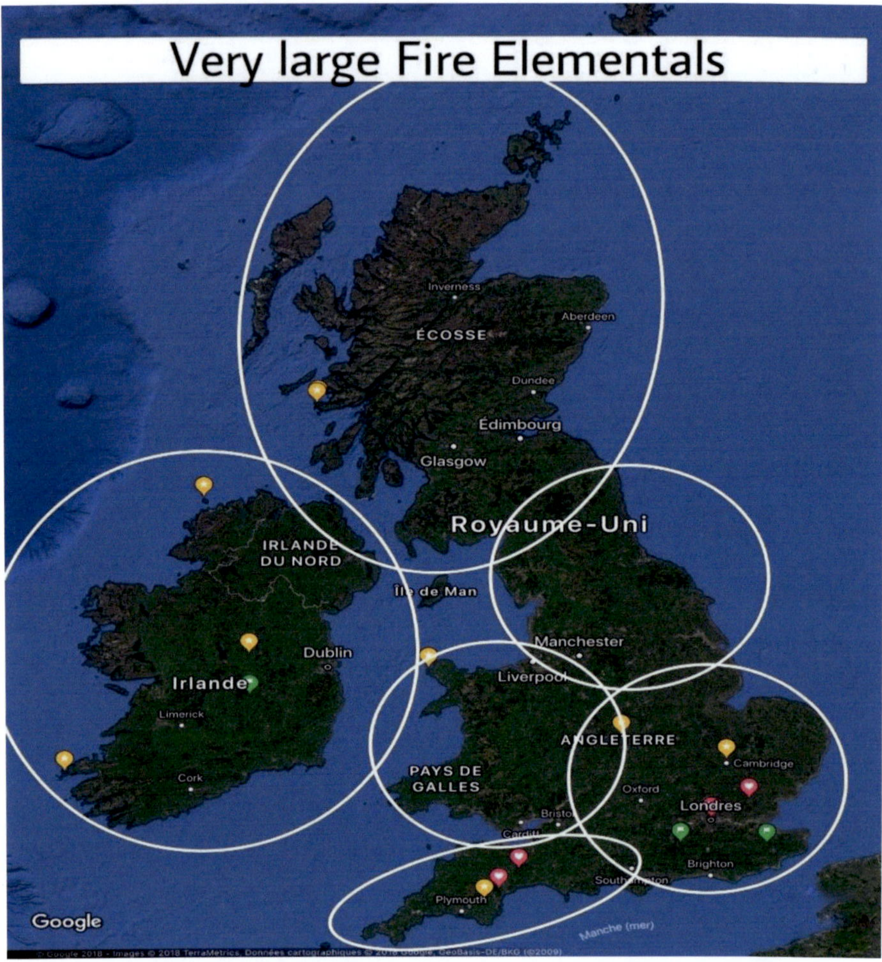

The fire elementals are in charge of electricity production, distribution, and consumption, fire, heating, combustion motors, fire weapons, explosives, nuclear energy, and light pollution at night time, etc.

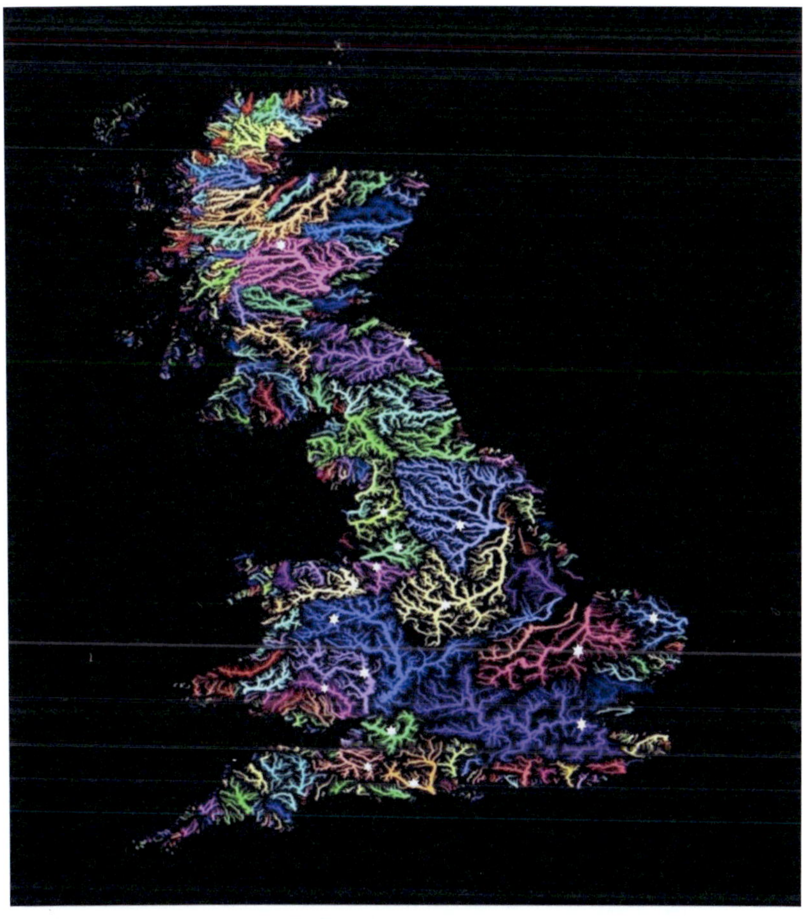

Britain's fluvial systems with their largest water elementals (see stars);
smaller rivers have smaller elementals not here shown. D. Perret

The white dots show the location of the Water elemental
of that River system. They can be sensed by means of
radiesthesia. Their position is stable. Map by Robert Szucs
See interview with the Loire Valley Elemental page 55.

River systems of Europe with some of their very large Water elementals' location. Map by Robert Szucs

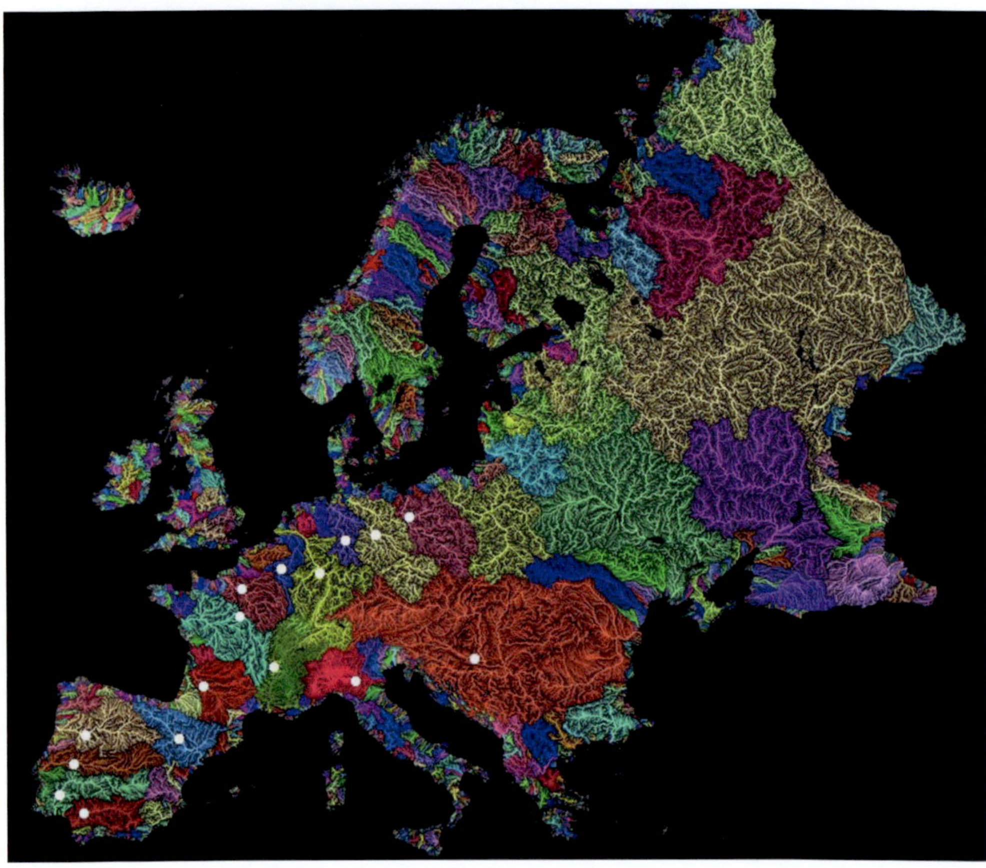

In the following pages I will present an important type of nature spirits: Devas.

The **Devas**

The Deva of a plant, flower or tree can move within all the four ethers and coordinate the efforts of all elementals. The elementals can only function within their own ether type and cannot go into other ethers. An earth elemental cannot for example communicate directly with elementals of the other elements (water, fire, air), nor can they see or perceive them. They communicate through the Deva.

The Devas are a more highly evolved species than the elementals. They form their own hierarchy. The coordination of the shape of a group of trees is made by the cooperation of the tree Devas. The highest Deva in a country is the Deva Queen. Larger countries have some-times several Queens, France has four Deva Queens.

There is only one Deva Empress on the Earth, she is on a level above the Queens. Below the level of the Queens, we find the many Deva Princesses. [20]

A tree Deva can be considered as being the spirit of the tree. We can communicate with them, and ask for example whether they are feeling well? whether we can do something for them? where to plant a young tree, etc.

Tree Devas coordinate all the work that is needed to make a tree grow and to maintain its health. An army of small elementals accomplish the work.

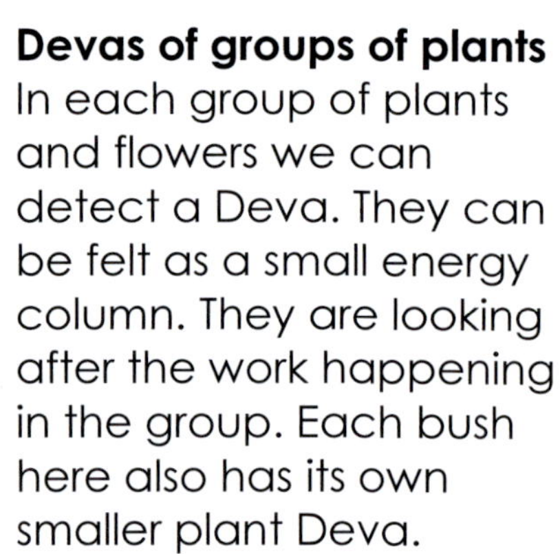

Devas of groups of plants

In each group of plants and flowers we can detect a Deva. They can be felt as a small energy column. They are looking after the work happening in the group. Each bush here also has its own smaller plant Deva.

Foto above: Deva of a small group of flowers

Devas of land parcelles

In this photograph the Deva of the water meadow is found on the ground beneath the hanging branch. In springtime the whole meadow is covered in yellow marsh marigolds and a little later with yellow irises.

The Organization of the Invisible Worlds
Universal Intelligence
The 'Father' impulse
Loving intelligence - electricity
The Angelic Hierarchies
Seraphim Cherubim Thrones Cupidos .... Earth Angels
Landscape Angels Angels or Spirits of Nations Angels or 'Spirits' of Towns, Countries, etc.
The Creation (visible and invisible)
Elementals Devas Dagdas Elves and many more
Very large water elementals of River systems
elementals of earth, fire, air, 5th type
Nature Spirits
The Earth 'Mother' impulse
Intelligent Love - magnetic forces

The celestial or **angelic hierarchies**

Besides the nature spirits, that we can call beings of Mother Earth, there are the angelic hierarchies of the Heavens. They bring down the impulses from Heaven or Universal Intelligence into Creation. The Creation process happens through the coming together of the heavenly (male polarity, electric) impulses with the (female polarity) energies of Mother Earth (the magnetic forces of Creation). The beings from the angelic hierarchies are the electrical forces or impulses of Creation.

See also list of the 21 spheres of the Divine Field in the appendix

Hierarchies in Nature

The various hierarchies in nature create a natural and harmonious organization and in nature there is no ego and no fear. The hierarchies that operate within nature – in a wide sense, including the divine field - are based on competence, wisdom, and experience. A being with more experience and wisdom will naturally be consulted by younger or less experienced colleagues.

After this brief introduction to the five elements, the five types of elementals, and other nature spirits we now come to the next building blocks of physical manifestation. These are the four types of energies we find in the etheric energy fields around any physical being and any plant or tree: the four ethers and the four layers of the etheric. This etheric brings us the understanding of how energy becomes matter.

The four ethers

They are the four formative forces working in the four etheric layers and that are derived from the original impulses: electricity and magnetism. The reflector ether is the first densification level of the electric force, the light ether a denser form of the same electric impulse. The angelic beings work in the reflector and light ethers.

The life ether being the first manifestation of the magnetic impulse or force, the chemical ether the more densified phase of the same impulse. Page 109: The air elementals work in the reflector ether, the fire elementals in the life ether and the water and earth elementals in the chemical ethers. The action of the chemical ether is to organize subtle energy into specific shapes.

The etheric   (see in-depth description in 3rd part)

The etheric energy field is the organization of the four ethers or types of creative forces into four layers: *chemical ether, light ether, life ether, reflector ether*

It exists around stones, plants, animals, human beings, and around the planet earth. The etheric is created and brought about by the divine beings of sphere 14 (Exusiai angels, see list at the end, p. 134). These angelic beings provide the constant energy supply to the reflector ether of an organism as described below. The reflector ether then transforms and brings the energy supply for the life ether, this is then transmitted to the light ether, which in turn provides the chemical ether of the organism with energy.

When this chain of energy supply is interrupted or too weak, the chemical ether cannot keep the cells alive, which ultimately leads to a collapse of the energy fields and the death of the organism.

The whole process of transformation of energy depends on many beings successfully cooperating and is thus a highly complex phenomenon. The sphere 14 beings are the life bringing force behind all of this. The Devas are the coordinators of the elementals.

The CO_2 is transformed by the air elementals of a tree, and its carbon C is brought to the chemical ether - through the pores of the leaf - where it helps in building the cells. The O oxygen is released into the air.

The nutrients in the earth are transformed by the gnomes (earth elementals) and brought to the chemical ether. The physical water is transformed by the undines and also used in the chemical ether. The fire elementals transform the heat and the rays of the sun into energy for the chemical ether.

The physical substances alone are not sufficient to keep any organism alive. All these 'physical' substances essentially bring energy and have been created by sphere 15 ('Dynameis' see page 135) beings to help physical life. In winter and also during times with less insects, the metabolism of a tree slows down.

Although I am neither a biologist nor a chemist, I will venture into the following subjects in order to explore what elementals do:

The elementals are the workers that on a molecular level make processes actually happen.

Our usual level of understanding doesn't explain in detail how a chemical process is happening. We limit ourselves to believing that the bringing together of two substances is enough to make the process happening as if putting flour, sugar, butter, and eggs together would be enough for making the cake or the carrying of screws, wires, steel, and rubber tires into a factory and closing the doors would create the car that comes out the other end.

"In plants the role of chlorophyl in photosynthesis is vital. Chlorophyl, which resides in the chloroplasts of plants, is the green pigment that is necessary in order for plants to convert carbon dioxide and water, using sunlight, into oxygen and glucose. Photosynthesis in plants can be compared with the digestive system in that they both break down vital elements to produce energy that is used for nourishment and growth. Some of this energy is used immediately, and some is stored for later use."
www.gardeningknowhow.com/specieal/children/photosynthesis-for-kids.htm

So far so good. But do we know how the processes described here actually happen on a molecular level? The whole process is happening on an energy level and is carried out by elementals.

We can try to imagine the hydraulic system of a tree that endeavors to bring water to its leaves, bark, branches, and roots. We might think that the evaporation of water by the leaves in the sun, or the capillary effect would be enough to help the liquids rise up to the tree top. Yet, we would be unable to reproduce such a complex hydraulic system guaranteeing the life of a tree. But liquids rise to the top even before there were any leaves and after they had left in autumn.

How do water elementals contribute to this? I am told they induce a process comparable to atmospheric pressure – liquids move from higher pressure to lower pressure – happening in the chemical ether. Water elementals know how to transform chemical ether energy into physical water particles (H_2O)…and back into energy. That knowledge is also being used in the moving of liquids within any plant and tree. With our limited understanding, that is as far as we can go until we learn more about the chemical ether.

We usually forget that it is the activity of a conscious intelligence that is directing all the required processes. Without such a consciousness, no manifestation is possible, as we saw. In the same way no machine, no computer, no matter how sophisticated they might be, would produce consciousness, feelings, or purpose.

Federico Faggin wrote: 'Physics currently describes a universe that is holistic and dynamic but deals only with exteriority: measurable events occurring in space and time. To explain the existence of consciousness and free-will, we need to address interiority as well (an inner feeling dimension to objects, DP).'

WIM, 11 FEBRUARY 2021, Faggin: One and the consciousness units [2]

Visible examples of non-physical influence in nature

In the left-hand photograph, we observe how a young chestnut tree grows towards the right, away from the etheric field of the large chestnut tree to its left. The sun and the south are on the right side of the photo, so it is not the shade of the big tree that forces the small one to bend to the right. The etheric energy field of the large tree is, also on this photo, easily perceivable with radiesthesia and dowsing tools for instance.

On the photograph on the right, we see two trees taking on a single global shape, as if pruned by human gardeners. This is an example of a group organization with the help of a common consciousness, here a group Deva.

Nature Healing

Does nature need healing or simply need to be left alone?

I will explore this question through the themes of clear-cuts in forests, reforestation, rewilding areas, and then describe the cooperation with nature and the concept of healing for nature spirits.

Working with spiritual energies means to draw on energies beyond ego and beyond any personal search for benefits or recognition. It links to the outer reflector ether and through this to the divine field and to the spiritual or quality aura. This is the area of a person's aura that contains their spiritual qualities that they brought into the physical dimension with them at birth. These belong to one or several spiritual colours or so-called rays.

The cooperation with nature spirits is of mutual benefit, and is absolutely necessary in this time of urgent environmental problems.

Rewilding of areas

Our poor understanding of the intelligence of nature shows when part of a landscape is designated to be rewilded. It is a mistake to simply leave the area without human interference and imagine that it would revert to a 'harmonious' natural state. This is not the case.

Human interference can completely destroy an eco-system to such an extent that the harmonious network of nature spirits is no longer operating. Various factors can prevent the return to a well working organization of nature spirits. Leaving a damaged ecosystem to itself can more often than not create chaos, and an unharmonious return of plant life which can show in different ways. Invasive or especially fast-growing plants may take over and prevent a balanced regrowth.

However, areas which still have a functioning organization of nature spirits, will eventually recreate a harmonious natural area. There can also be people with special natural skills ('green thumbs'), that can lead a rewilding into the right direction. Communication with the local landscape angel or regional Deva may also be very helpful.

Nature is a complex yet vulnerable system. There is a wise hierarchy of beings working together in order to keep a harmonious balance. See Appendix p 123 the Yellowstone Park example, reintroducing wolves.

Clear-cuts in forests

The practice of clear cutting whole sections of forest and leaving no young or old 'mother' trees is very damaging and creates a similar unbalance. The whole ecosystem, the community of trees, bushes, undergrowth is destroyed, even more so when huge machines operate. In wet weather these new machines plough very deep furrows into the soil so all the seeds and shoots are buried and dead. It is then practically impossible for a harmonious ecosystem to regrow. Often fast-growing species such as, here in Dordogne, sweet chestnuts or brambles, will eventually take over and not allow any other species to grow.

Reforestation

Replanting trees in a vast area, to counter climate change meets the same problems if it is carried out without the wisdom of nature. Monoculture and planting in rows, does not create a balanced ecosystem.

Natural disasters and ecocatastrophes

As awareness of the different causes grows, it has become quite difficult to differentiate between natural disasters, man-made and man-accelerated upheavals in nature. The Californian Forest fires of 2020 have brought to evidence that ancestral habits of periodically clearing undergrowth were a very good prevention of wide spread wild fires. This had been forgotten and neglected for many years. In 2021 numerous wild fires in Western Canada happened because of human neglect, and many were caused by lightning. In 2001 the Flensburger Hefte published a book of interviews with nature spirits, now translated into English. [4a]

Wolfgang Weirauch interviews in it the 'Fiery One', which is a great fire elemental. "Why are there more and more thunderstorms recently especially in this year (2001 presumably, or 2000 depending on the delay between interview and publishing date)?" …The 'Fiery One': "Amongst other things, it's time because the stars are standing in particular constellations (for instance fiery Mars and Sun in fire signs Aries, Leo or Sagittarius, DP), because humans are behaving very badly and because the middle and upper hierarchies are sending their intentions down to the earth in the form of strong impulses. … A thunder-storm is caused for example, when groups of people try to exploit other groups of people, exploiting them on an economic level. Unfair trade generates a lot of thunderstorms."

These examples suggest that there may be deep and multiple causes behind natural catastrophes and the imbalance seen in nature. Not only human actions but also human thoughts can be part of the complex background of causes.

I am thinking about natural water retention areas of floodplains, that act as natural storage reservoirs. They allow large volumes of water to be stored in case of flooding, and then to be slowly and safely released down river and into the groundwater.* These are part of nature's intelligence and could perhaps prevent devastating floods, such as those in Belgium and Western Germany in July 2021. In this case climate change caused a much larger quantity of water to be drawn up into clouds. Yet, why the rain came down in a relatively small area is likely due to another factor that needs to be researched.
*) partly drawn from wwf.eu Danube floods Natural retention areas needed

In an interview with nature spirits ([4a]) it is hinted that earth quakes often take place in especially under-privileged areas of the world, that then suddenly get more attention through the event: Aquila 2009 (Italy), Kashmir 2005 (Pakistan), Bam 2003 (Iran), Haiti 2010.

The intelligence behind Nature brings a number of very interesting and surprising new insights. It gives us a glimpse of the widest definition of 'nature', having us look at the 'Intelligence behind the *World*' or 'behind *Creation*' as a whole.

Structural rethinking

A lot of structural rethinking needs to be done in the near future in order to rectify devastating past mistakes concerning inappropriate, environmentally damaging agriculture, unbalanced use of water, to name only a few. The cooperation with the wisdom of nature is inevitable because the problems have become so complex. This is due to our decades, if not centuries long, loss of contact with nature's wisdom . For centuries we thought we were going to reinvent the egg and now we see we might have to ask the chicken.

Cooperation with nature

We need to reflect on how we can contribute to a harmonious coexistence with all beings, visible or not, with whom we share this Earth. There is no other way than cooperating with nature, that is with the intelligent beings behind nature. This means becoming conscious of them, respecting them, and asking them how they are and what we can do for them. Whether we are personally able to get a clear answer from them or not is secondary. Being conscious of them and asking them means already showing our respect, our care.

When I was meditating with friends in Switzerland, we had a quartz crystal in our middle of the room. At a certain moment we all sensed a line coming towards the crystal. When began to ask questions, a landscape angel near Mühleberg nuclear power station told us that the area needed healing. This was because pollution had affected the soil around the power station to a depth of more than 300 meters and that no one had noticed this. At first, we did not know what we could contribute. We then decided to meditate with our awareness in our hearts and beneath our feet which we imagined to be deeply rooted in the earth. We then connected with the area around the power station. After a relatively short time the line coming towards the crystal disappeared and the landscape angel told us that our meditation had helped.

During a training group, all the course members noticed that a new type of signature showed up at the crystal. It consisted of two lines forming a 30° segment. Using the Yes/No question technique with the Hartmann Antenna we asked who it was. It was a large fire elemental

located in an ancient unused church in the middle of

Chartres. He explained that his work was affected by people's emotions about a new electricity meter (LINKY) that the electricity company was gradually equipping households with. He had no particular opinion about the device itself, but said that the very emotional public debate was affecting his task of distributing energy, as the electric wires were, along with the electric current, also transporting information (from that meter for instance).

In November 2017 I contacted the Nore river Elemental in South East Ireland. The elemental is located at its most northern bend near Bishopswood where the river starts its southward journey. The **River Nore** (Irish: *An Fheoir*) is one of the principal rivers in the South-East Region of Ireland. The 140-km-long river drains approximately 2,530 square kilometers of Leinster and Munster. The river is home to the only known existent population of the critically endangered Nore freshwater pearl mussel, and much of its length is listed as a Special Area of Conservation. The elemental specified that the Abbeyleix water authorities had been putting too many chemicals into the domestic water system for the last two years. Many of chemicals ended up flowing into the river Nore. These elementals are in charge of the whole water element installations in their area, not just the river itself.

On several occasions the water elementals of entire river systems of the Dordogne, Lot and Loire contacted me. Abundant rainfall had brought unusual amounts of earth into their rivers and they needed help to deal with it. As I understand it, distant healing helped connect them with some beings of the Divine Field, who they seem unable to contact themselves. They always reported afterwards that the distant healing had worked.

The numerous contacts and explanations I received over the years have made me aware of the many daily tasks that elementals have to deal with, that we are usually unaware of.

On two occasions Devas of different sections of forest contacted me about a deterioration of their areas. Once it was because of too many badgers and another time because of too many stags. Contact with the group souls of these animals helped to rebalance the situation. The problem with the badgers had happened because too many offspring had remained in the section where their parents had settled. The youngsters needed to find their own territory.

I have been sending distant healing for many years. Distant healing is usually directed towards humans who have asked for it.

I have described above how, some years ago, nature spirits began to contact me and requested healing, telling me that prayer would be helpful. Afterwards they let me know that this was what they had needed. I wondered what was happening and what we actually can contribute that they could not do themselves?

I realized that being physical we can do things they

cannot, for instance sending a prayer using our thoughts and feelings. We seem to have a contact to the spiritual

realm that is particular to human beings. Although scientific research has shown the positive effects of prayer, it is obvious to me that we still know very little of the power of our thoughts and prayers. I came to understand that perhaps, because we are physical, we can function like a telephone operator from the past. We can facilitate connections between nature spirits and beings with a required knowledge in the divine field, that nature spirits could not directly contact for some reason.

There are more aspects that became evident with this healing process for nature: Over the years nature spirits have explained how they function, the kinds of problems they are facing and how they are organized. [16]

Deep healing happens when spiritual energies from the universal intelligence are allowed to take over. That is a consciousness accessible to us and coming from beyond the ego – in other words: from the sincerity of our hearts. We all can do this.

The animal kingdom
Many initiatives for the rights of nature are concerned with geographically limited areas. In my observation

most intelligences in nature are organised geographic-ally as well. The animal kingdom is organised differently, according to the species. The appropriate partners to communicate with animals would often be through the group, tribe, flock or herd souls (of deer, wild boars, salmon, bees, storks, etc.).

Harmony in an ecosystem means respecting all sentient beings. We need to consider that other types of beings may also be organised and possibly need to be addressed differently.

We need to discover ways of communicating with river systems, landscapes, trees, and animals, in order to move away from an exclusively anthropocentric approach. Many first nations throughout the world have kept this deep wisdom about nature alive in their cultures.

The Spirit of the Valley never dies.

This is called the mysterious female.

She is the Virgin Blackness,

The Black Virgin and Earth Mother.

The gateway of the mysterious female

Is called the root of heaven and earth.

Dimly visible, it seems as if it were there,

Yet use will never drain it.

Inspired by
Tao Te King, Lao Tzu, chapter 6

3rd Part

An energy-science view on how consciousness becomes physical manifestation

In this section I will present some thoughts on the transformation of consciousness or energy as it moves into physical manifestation. I am drawing on texts on the science of spiritual healing as taught by Bob Moore, on information from 'C' as well as on my own observations.

The etheric is where the transformation from invisible energy into physical matter happens. This is central and if we want to understand how consciousness becomes physically manifest, we need to study the etheric.

Discovering of the etheric
My discovery of the etheric was inspired by various sources. Bob Moore often taught about the characteristics of the four ethers and the five elements. I clearly remember that each time this subject was mentioned I pricked up my ears and wrote down carefully what he

said, although it was only a few sentences here and there. It took me a very long time, in fact decades, to familiarize myself with what the function of the different ethers might be. The quotation from Bob Moore (page 59) about the reflector ether being the highest energy contact we could attain particularly moved me. See also page 151

When I teach about spiritual healing and the layers of the aura my explanations are based on my own perception of each energy field and the distance it would show up from the skin surface. In 2003, I discovered that the etheric had expanded considerably since the time that Bob Moore had been teaching (1979-99). I measured that the outer edge of the reflector ether was now about 64 cm from the skin surface, and that the edge of the life ether was 12 cm from the skin surface which was about the former distance of the reflector ether. This widening of the layers made it now possible to distinguish three sublayers in the reflector ether and two in the life ether. With the help of 'C' I received explanation about their different functions.

It is my wish to explain what the precise work of nature spirits that prompts me to write in detail about the etheric. My hope is to increase the understanding of why it is necessary and useful to include invisible intelligences in the process of establishing legal rights for nature.

The more detail I went into in my exploration of the etheric, the more aspects I found there were to understand about how energy is trans-formed into physical manifestation. One answer always led me to the next question. I realized once more how long it takes to turn information into felt knowledge, and although I am far from a complete understanding, my hope is that in writing this I may help deepen the love and respect for nature and the wonders of Creation. Obviously, we need to expand and completely review our view of biology and life in general.

The Etheric: *Consciousness becomes manifestation*
The four layers of the etheric are probably very close to what quantum physicists would label four different quantum fields, thus four quite different worlds and systems. In this section (and in the section on the etheric in the appendix) I will quote extensively from Bob Moore who was a rare expert on the etheric.

The etheric is an intersection of different energies from very fast-moving energy, such as intuition, higher consciousness energy and the slower energy such as the physical body or the energy of the astral aura (energy field of emotions and feelings) and mental aura (energy field of thoughts, intuitions, higher concepts).

All experience – in terms of perceptions through our five senses - is stored in the etheric. The function of the etheric is to indiscriminately store energy that can then be used for memory or for various activities, like running.

The four layers of the etheric can be found surrounding and penetrating any living organism, the whole earth, plants, trees, animals and to some extent around any inanimate object. One layer can be more predominant around certain plants or places. Objects don't seem to have a reflector ether but do have a life ether surrounding them which extends up to 12 cm.

The four layers of the etheric:
Chemical ether, light ether, life ether and reflector ether

The two first layers are inside the physical body and are separated from the two outer layers at the level of the physical skin surface. You can easily learn to sense the two outer ones as they are organised like a layer cake. As I mentioned the reflector ether extends to about 64 cm, the life ether to about 12 cm from the skin surface. With human beings the expansion at a particular area of the body depends on how relaxed the person is.

REFLECTOR ETHER
- outer layer*
- middle layer
- inner layer

LIFE ETHER
- outer layer
- inner layer

skin surface
- -

LIGHT ETHER / CHEMICAL ETHER

*) this layer varies from 5-15 cm depending on the person

Even if the etheric of all physical beings is structured the same way, there is a great difference in the use of the etheric when we come to human beings. We can use our mind either to help or hinder the natural functions of the etheric (mainly through our emotions.) Other physical beings cannot do this. I have put more information about aspects that exclusively concern human beings in the appendix. In the following section I have included quotations that are of general significance to all physical beings. When we observe characteristics of the human etheric, we gain more understanding of the general performance of the etheric.

The etheric is often compared to the water element because of its particularly fluid energy movement and energy distribution. The human etheric has etheric energy streams (the eight psychic streams that Bob Moore used in his teaching, Chinese acupuncture meridians, Indian nadis, etc.) that operate an overall distribution of the energy.

Similar energy streams can be found in a flower, a plant, a tree and in a landscape. In plant life the energy lines would obviously follow twigs, branches and the stem in a tree. Trees also have energy centres, comparable to the human chakras. The tree only has two, one in the root area and another two fifths up the trunk, the exact place where the tree Deva is usually found.

"The etheric is a storehouse, a storage of energy. What that energy is, has come rather loosely under the name of **prana**. …prana is a word that comes from India and means life force. That life force is something that you and I need, but in needing it, our distribution of it is very important and so that distribution is reflected in the movement of energy from some states of consciousness into the storage that we have etherically and that is operating through the chakras. In operating through the chakras this combination of energy that we have, is then reflected into the nervous system." Bob Moore

Plant life also has a type of nervous system.

The etheric has the densest energy of all our energy fields and vibrates at a much lower frequency than the astral, mental or spiritual layers (drawing p. 126).

Chemical ether

This is the layer that is found deepest inside the body and is linked to the element of water, to sound, colour and numbers. The link with numbers can be seen in the periodic table of the chemical elements (Mendeleyev). The chemical ether is responsible for the nourishment of the bones, muscles, cells and organs. It is linked to the liver, the spleen, metabolism, to the assimilation and elimination of substances not used in digestion. Its symbol is the lower half of a circle or the half-moon.

"The chemical ether structure is basically **colour**. It's embodied in all the ether atom structure that surrounds you and so that is why you need to revive the whole activity, particularly when you are using a lot of the storage of energy in the etheric ... but it has got to be revived in a cycle of harmony or a cycle of rhythm."

Light ether

"This ether is also called the truth ether and is strongly linked to the element of air as well as to universal structures like light, love and life. Light is essential for healing depression and more generally in the maintenance of life. The light ether is found just under the skin and needs to be vibrant and have a good pulsation. With humans it is very present around the head and especially in the zones linked to thinking. Its symbol is the triangle."

Life ether

This ether is linked to the element of earth and to everyday life. It extends about 12 cm out from the skin surface. It feels denser than the reflector ether. Its symbol is the square.

Reflector ether or warmth ether

This layer of the etheric is linked to the element of fire and is found furthest away from the skin surface. Its symbol is the circle.

(more about the four layers of the etheric on page 140)

"The etheric structure was already in existence before your physical body was structured. The reality is, it is not the etheric that is a counterpart of the physical, but the physical is a counterpart of the etheric.

The use of these **ethers** that we see in the etheric are not just in our etheric, they are also in the etheric around the world and this is why, when we look at the development of any one person, it can never be a singular development, it has got to relate to other people, it has got to relate to the universe, because we have all these compositions of ethers that are reflected around individuals or reflected around the world, or, I would assume, are also reflected around other planets. The whole composition of creation then is one that is showing its connection to structures that have formed a means to have things created, which the etheric does.

The ethers, being part of the composition of the etheric can be affected or be not allowed to utilize their movement because of a distortion-situation that we would create, and so in that way sometimes we are not drawing on the balance of the four states of the etheric.

The movement of energy that we find related to the **nervous system** itself also relates to what we see or sense or feel etherically. We are using **this word 'etheric'** and there is no other way, that I know of, that we can start to compare energy, as we are looking at energy in its movement, except we do it through the etheric."
Bob Moore.
(Bob Moore refers here to the very loose use of the word 'energy' which remains basically a mystery as it includes all kinds of 'worlds and quantum fields' and the many different types of energy we actually know very little about.)

"The etheric structure is a counterpart of the physical. Being a counterpart of the physical, this means that we have some areas of comparing energy movement with what we see in the physical body itself. That energy movement is reflected in the changes that are continuously going on physically, for instant in a joint. The change of movement in a joint is a change of movement reflecting energy, not only nervous energy that we see in the physical body, but also energy that we see related to the etheric structure."

We will now have a closer look at how consciousness/energy becomes physical manifestation. It is essentially the four layers of the etheric that are involved in this, together with the beings working in those layers.

4 etheric layers

angelic beings of sphere 14 are working in the outer layer of the reflector **ether** together with the air elementals

fire elementals working in the **life ether**

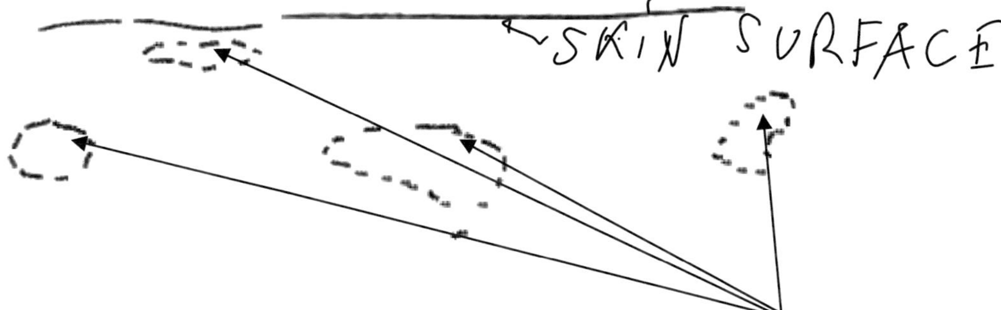

earth and water elementals working in the **chemical ether**; angelic beings of sphere 15 are working in the **light ether**

The Creation Process

Father Heaven - Universal Intelligence
electric impulse

↓

Beings of the Divine Field

↓

Reflector ether - 'Powers' maintaining the
overall balance and conditions

↓

Light ether - 'Virtues' creating light beam
structures

↓

Formation of the physical on the level of the
Chemical ether – water & earth elementals

↑

Nature Spirits - Devas

↑

Mother Earth - Universal Love
magnetism

All manifestation of consciousness, starting at the level of
Universal Intelligence/God, can be understood as a
gradual yet complex transformation downwards of
frequencies through a succession of energy/quantum
fields to reach the slow vibration of physical matter.

The steps: manifestation originates from the original impulses coming into the upper reflector ether downwards into matter. We need to remember two principles:

1. Principle: All Manifestation, visible or invisible, originates from a conscious sentient being. as mentioned, p. 60:

and p. 62: 6. Principle: The Universal Purpose is to give individual form to creation and experience it.
... the Father/Mother/Consciousness is in equilibrium within the Universal Consciousness/Awareness where 'Father Consciousness' is Universal Intelligence/using electricity and 'Mother Consciousness' is Universal Love using magnetism.

Beings of the angelic hierarchies or Divine Field work in the reflector ether and in the light ethers:
The Exusiai - 'Powers' (sphere 14*) : are working in the reflector ether (of plants e.g.) maintaining the overall balance and conditions needed for the Creation process to take place. *) list p. 135

Dynameis - 'Virtues' (sphere 15) are working for example in the light ether of plants by creating light beam structures (grids, lines, platonic volumes) that then become the precise construction plans for the earth- and water elementals to build the cells within the chemical ether, the densest form of the ethers, closest to physical matter.

The beings from the angelic hierarchies are part of the electrical forces of creation. (p. 62)

On the level of the chemical ether, it is the undines (small water elementals) that organize the *magnetic forces* that hold together the atom and create the components of the atom nucleus: the neutrons and protons. The gnomes (small earth-elementals) are handling the *electric forces*, creating the electrons circling the nucleus. The plan of how many electrons, neutrons and protons need to be assembled, as described in the periodic table of Mendeleyev, is part of the structure present in the light ether along with the necessary order of energy lines to create a particular leaf, etc.

We must remember, that, on an atomic level, everything is energy – nothing is solid. What we consider to be physical matter is a mental/optical illusion we have become used to live by. What is happening in the chemical ether in terms of Creation is the structuring of (magnetic and electric) energy in a particular way that gives us the illusion of a solid physical world.

Maintaining and restoring of life
The steps for the maintenance and restoration of life, are similar ; the same beings are continuously at work in the three lower ethers to keep a flower, a tree, and a plant alive. I have described (page 54)how the Devas are the beings that assist in the daily operation of maintaining and protecting life in trees and plants.

The role of the light ether in the forming of cells

The photo-montage on the left of light ether lines on young wisteria leaves, was felt with a radiesthesia or dowsing tool and energy perception by the author. The same lines are then found as veins in the finished leaf.

Photo on the right: full grown wisteria leaf.

In these two pictures we can see the two last phases of the physical manifestation. The light ether beings create the back bone of the structure which is the detailed construction plan along which directives the beings operating in the chemical ether are building the cells. We can see the same structures in the left photograph in the light ether as we see in the right hand photograph as the veins of the full-grown leaf.

Prior to this construction phase we have the blue print plans coming in through the outer reflector ether (the interface with the 'universal intelligence'); from there the blue prints move into the lower reflector ether (home of the blue print of the etheric) and from there directly into the light ether. The life ether of plants comes into action mainly after the construction phase, once the plant is physically operational and alive.

Comparable to an architect designing a house, the impulse or original design comes through the reflector ether with the 'Virtues' (sphere 15 beings of the Divine Field). They carry the blue print design into the light ether, which is inside the physical 'body' and then create the structure of energy lines (see photo montage of a wisteria leaf). The Virtues are the architects, the intelligent invisible beings originating in what we can call the Divine Field or Field of Universal Principles.

In its uppermost layer the reflector ether is an interface with universal wisdom (including universal laws, higher beings, major or very large nature spirits). With the help of the chemical ether the components of matter begin to form around the structures imprinted in the light ether.

Plant seeds are the beginning or starting point for the light ether blue prints. They carry the genetic code which is the anchoring help for the blue print details of the light ether. All the details of the blue print plan are already in the light ether. The **genetic code** details operate exclusively in the light ether. They are created and changed by sphere 11 beings (Archangel Hesediel – Divine Will); these pass on the instructions to change the genetic code to the gnomes (small earth elementals) working in the chemical ether.

Federico Faggin: "Cells are quantum-classical systems, not classical system as we have been told. Their essential interdependence with the environment is another clue that they have not lost their individual connection with the wholeness of the quantum fields, that's why I think living cells are conscious, though their consciousness is likely very different from our own."

WIM 11 DECEMBER 2020, FAGGIN *Consciousness is fundamental* **2**

I have described the creation process with a wisteria leaf, and this Creation process is a basic pattern for Creation through the etheric for all physical beings: plants, animals, human beings.
We need to have a closer look now at how energy from one layer gets transformed into the next type of energy or layer.

Transformation of frequencies: from one layer of the etheric down to the next denser and slower vibrating layer

**The reflector ether of large trees –
and the transformation of its frequency through sound**
A large tree can have a very big reflector ether exten-ding way beyond the diameter of its crown, especially in summer. It is significantly smaller in winter. It is the tree's energy storehouse. That reflector ether energy is produced by the transformation process from CO_2 to Oxygen and Carbon. The transformation from the higher reflector ether into the slower vibrating energy of the life ether is done – according to 'C' – by the humming sound

of the flying insects around the tree. The sound is all on one frequency, and one can easily hear it. I noticed it to be constantly at 256 Hz or C when tuning A at 432 Hz; 256 Hertz seems to correspond with the frequency of the life ether. For more details see p 153

In the photomontage we see the outer rim of the reflector ether (white band) as it was then in the beginning of June 2021. The black line would be the position of the reflector ether in winter.

The transformation down from the life ether frequency into the light ether frequency also happens through

sound, this time the sounds of very small non-flying insects operating on the surface of the leaves.
The transformation of the light ether frequency down to the frequency of the chemical ether, in order to nourish it, is done by the sound of bacteria. Should it be true that glyphosate and such chemicals are destroying the bacteria in the soil, then the nourishing of the chemical ether through the sound of bacteria is in danger.

The reflector ether of plants and flowers
Plants and flowers receive their energy from water, earth, nutrients, and the sun into their reflector ether. The transformation of the reflector ether energies happens through the sound of flying insects, of crawling insects for the transfer from the life ether into the light ether, and the sound of bacteria for the third transformation into the frequency of the chemical ether.

The reflector ether of animals
They receive their reflector ether energy from food and water. All frequency transformation seems to happen through sound, for cats through purring, for dogs through the sounds of their breathing while sleeping, birds through the vibration (sound) of their emotions. The third transformation into the chemical ether happens with the sounds created by bacteria inside their bodies.

The reflector ether with human beings and the transformation of its frequency

According to 'C' the transformation down into the life ether happens through the sound of our own voice, the transformation down into the light ether through the sounds of our own thoughts and emotions.

The transformation from the frequency of the light ether to the chemical ether happens through the sound of bacteria in our body. Sound is the tool transporting the nourishing substance that comes with colour.

In our daily lives, when we suppress or cut off our feelings connected to an experience we have had – so that we do not allow it to get 'under our skin' – this creates a division between what has happened in reality and how we wish things to be like. The light ether is strongly linked to truth, we create a distortion. It creates a barrier on the level of our skin surface – where the life and light ethers meet and this hinders the transformation and nourishment of the light ether and ultimately the chemical ether and the cells it is supposed to nourish.

The human reflector ether is also our storehouse of energy. It is replenished by the prana energy we receive from food, drink, and sunshine, and also from the negative ions in the air. These are received first into the reflector ether then absorbed through the spleen and is then distributed to the main chakras: root, solar plexus, heart, pineal. The hara receives prana energy via the root chakra, the thyroid via the heart chakra.

In plants, trees and generally in eco systems, it can be pollution, illnesses due to unbalances (drought, excess humidity, air, earth, and water pollution through chemicals, poisons, etc.) that can create obstacles for higher energies to move into the chemical ether in order to keep plant life healthy. In such cases the elementals have much additional work to do to restore harmony. We may think that nature simply rebalances itself, yet it is the elementals that are accomplishing these tasks.

Over the years, I have been contacted by a variety of nature spirits who were asking for healing. The essence of spiritual healing is that higher frequency (spiritual) energy may be able to penetrate denser layers of energy. At first when the nature spirits requested help, I had no idea what I could do and how my intervention could work practically. Yet, I could observe that the use of the higher energy of prayer – they often asked explicitly for that – seemed to accomplish what they had asked for.

In the following example I came to understand how the thoughts and emotions of people had interrupted an energy connection between a landscape angel and a Deva of an area of land. This happened because two neighboring owners did not get on. One owner built a massive earth mound between the two properties. This was only the physical part of the mental-emotional wall between them. The connection of the Deva with the landscape angel seemed to depend on a direct line between the Deva and the angel being. That line was blocked by the earth/emotional/thought wall. Using a simple prayer seemed to bring in higher frequency energies (and beings) which repaired the energy connection.

See page 140 for further description of the etheric.

Spiritual impulses in humans

Humans receive impulses from different levels. The obvious ones being impulses from other people, intellectuals, artists, great statesmen, through observation of our surroundings, animals, nature, beings in need.

As a musician I have been mainly improvising for the last 30-40 years. I asked myself, some time ago, where did the ideas or impulses come from for 'my' improvisations? It took me some time to explore and realize that there were, beyond the ego level, three main areas where impulses come from: upper astral, upper mental and the spiritual level. Together with 'C' we worked out the different levels and spheres in great detail. See the three respective lists in the appendix.

When it comes to spiritual impulses one would usually employ the term 'spiritual' to mean that the impulse comes from beyond the ego, beyond selfish desires that originate from wanting power, rewards, money, recognition, to hide inferiority complexes, insecurity, and suchlike. Spiritual impulses originate from beings of the divine field. I have made an attempt to define this field through my own research; see the list in the appendix with its 21 spheres. These impulses can come directly from such a being to us or through the intermediary of the beauty in nature, in art, or the echoing of the impulse through another person. The quality of such impulses would be felt as uplifting, joyous, encouraging, generous and unintrusive.

As with plants and animals, these impulses first come into our reflector ether and from there into our heart as feelings or into our brain as images and words. When improvising music, I feel inspirational impulses as a sudden yet subtle 'wind under my wings' which allows me to become fully immersed in and get carried away in the experience of the sound clouds of the harp strings.

The spiritual origin of an impulse is one aspect, its watering down by influences from the human ego is the other. Plants, trees, rivers don't have this interference. This is one reason why being in nature is so beneficial for us. The spiritual impulse is unfolding more purely.

Appendix

The example of Yellowstone National Park

25 years after returning to Yellowstone Park wolves have helped restabilize the ecosystem.

Let us have a brief look at the reintroduction of wolves in Yellowstone Park. The ecosystem of the park had been greatly impacted by the killing of the last wolf in 1926. The reintroduction of wolves allowed the ecosystem to recover its natural balance, meaning that the intelligence of that whole ecosystem was allowed to operate again without a major human hindrance. We could call this the intelligence of the ecosystem, the Spirit of Yellowstone Park, yet it seems that it would describe itself as being a landscape angel.

(from my personal research: March 12th 2021: A brief communication with the landscape angel in charge of the wider Yellowstone Park ecosystem showed that he wanted the readers to know that he existed and where his exact location was: RJX2+X4 Devils Den, Wyoming, United States - (44.8498804, -110.3997381).

"The 1995/1996 reintroduction of grey wolves (Canis lupus) into Yellowstone National Park after a 70-year absence has allowed for studies of tri-trophic cascades involving wolves, elk (Cervus elaphus), and plant species such as aspen (Populus tremuloides), cottonwoods (Populus spp.), and willows (Salix spp.)

In Yellowstone National Park, wolves (Canis lupus) were extirpated from the park by the mid-1920s, absent for a period of seven decades, and reintroduced in the winters of 1995/1996. The historical presence, then absence, and now again presence of wolves in Yellowstone National Park (YNP) represents a natural experiment through time and an opportunity to study cascading trophic interactions. During the seven-decade wolf-free period, the collapse of a tri-trophic cascade allowed elk (Cervus elaphus) to significantly impact wildlife habitat, soils, and woody plants. For example, species such as aspen (Populus tremuloides) and willows (Salix spp.) were generally unable to successfully recruit young stems into the overstory on Yellowstone's northern winter ranges, except in fenced exclosures (Grimm, 1939 [30], Lovaas, 1970 [31], NRC, 2002 [32], Barmore, 2003 [33]).

Synthesis results generally indicate that the reintroduction of wolves restored a trophic cascade with woody browse species growing taller and canopy cover increasing in some, but not all places. After wolf reintroduction, elk populations decreased, but both beaver (Caster canadensis) and bison (Bison bison) numbers increased, possibly due to the increase in available woody plants and herbaceous forage resulting from less competition with elk. Trophic cascades research during the first 15 years after wolf reintroduction indicated substantial initial effects on both plants and animals, but northern Yellowstone still appears to be in the early stages of ecosystem recovery. In ecosystems where wolves have been displaced or locally extirpated, their reintroduction may represent a particularly effective approach for passive restoration."

Ripple, W.J., Beschta, R.L. *Trophic cascades in Yellowstone* [34]

The presence of wolves also changed the rivers. After their reintroduction it was observed that the erosion of the river banks has diminished causing the rivers to serpent less, the canals have deepened and little ponds have formed. Wolves are major predators having beneficial effects on a wide scale upon entire ecosystems, including the structure of rivers!

The reintroduction of grey wolves in the Yellowstone National Park is thus symbolically and ecologically some of the most important acts of fauna conservation of the 20th century. Once again Yellowstone is home for all the indigenous species of great carnivores: grizzly bears, brown bears, mountain lions and wolves. [35]

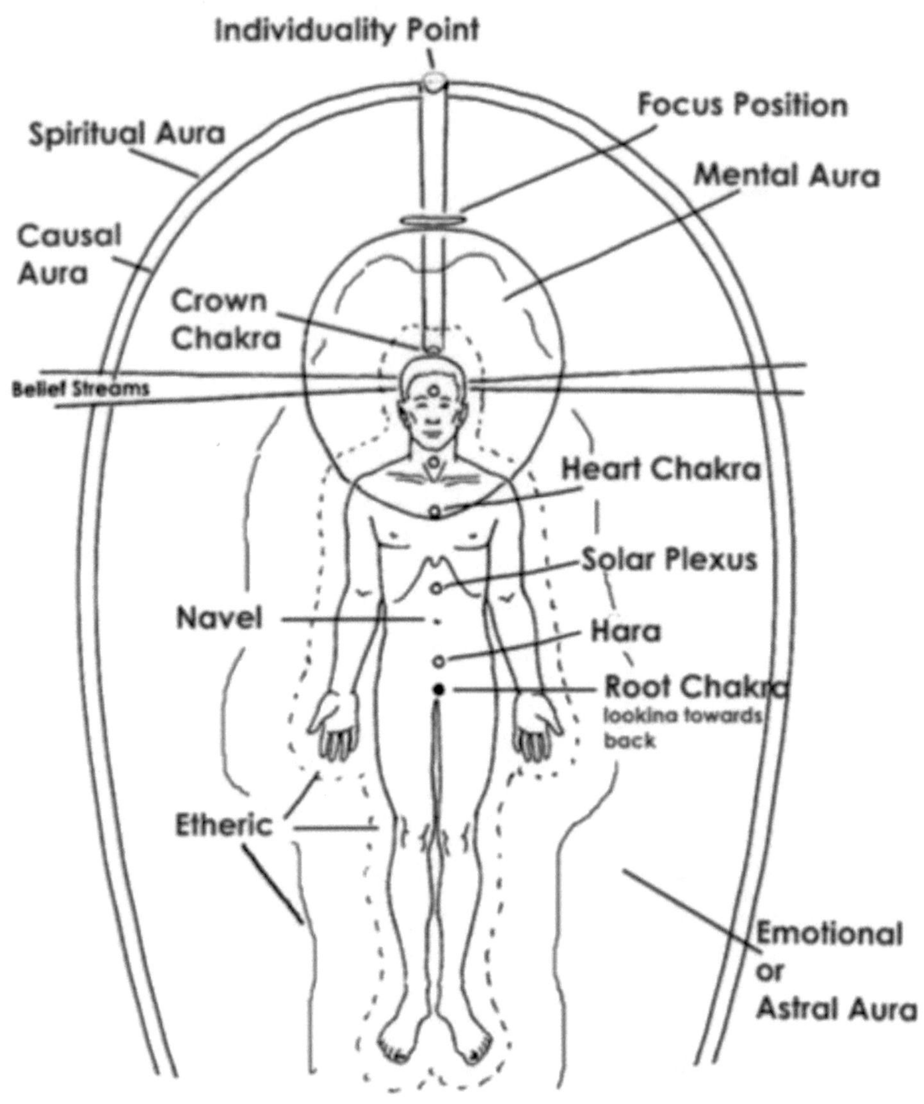

*) For more explanations see my book 'The Science of Spiritual Healing'[3]

Some layers of the human aura – their functions and place

Layer of energy	Function	Distance from skin surface	Existing since	Etheric
Etheric in the spine	Central current of cosmic-telluric energy or father/mother energy	Inside the physical body	conception	
Chemical ether	Nourishing the cells		3rd day after conception	
inner light ether	Truth, true feelings of events		conception	
Outer light ether	brings prana energy		conception	Etheric
Inner life ether	Perception, our 5 senses	0-4 cm	2nd month after conception	
Outer life ether	Memory	4-12 cm	2nd month after conception	
Inner reflector ether	Etheric blue print	12-17 cm	conception	
Middle reflector ether	a.o. is the astral blue print	17-26 cm	4th month of pregnancy and fulfils this function until age 14	
Outer reflector ether	a.o. the mental blue print highest energy contact interface with Universe, Nature	26-60 cm	Since birth and fulfils this function until age 21	
astral	emotional experiences	60-105 cm ca.	Formed since 4th month of pregnancy and is finished by the age of 14	
mental	Mental experiences	Essentially around head	Formed since birth and is finished by the age of 21	
Individuality point on the spiritual aura	Interface between timeless soul and this incarnation	some 105 cm	The decision of incarnating again, before conception	
Inner temporary soul layer	To form the etheric blue print	105-130 cm	The decision of incarnating again, before conception	3 temporary soul layers concerning this present life
Middle temporary soul layer	To form the astral blue print	130-155 cm	The decision of incarnating again, before conception	
Exterior temporary soul layer	To form the mental blue print	155-180 cm	The decision of incarnating again, before conception	

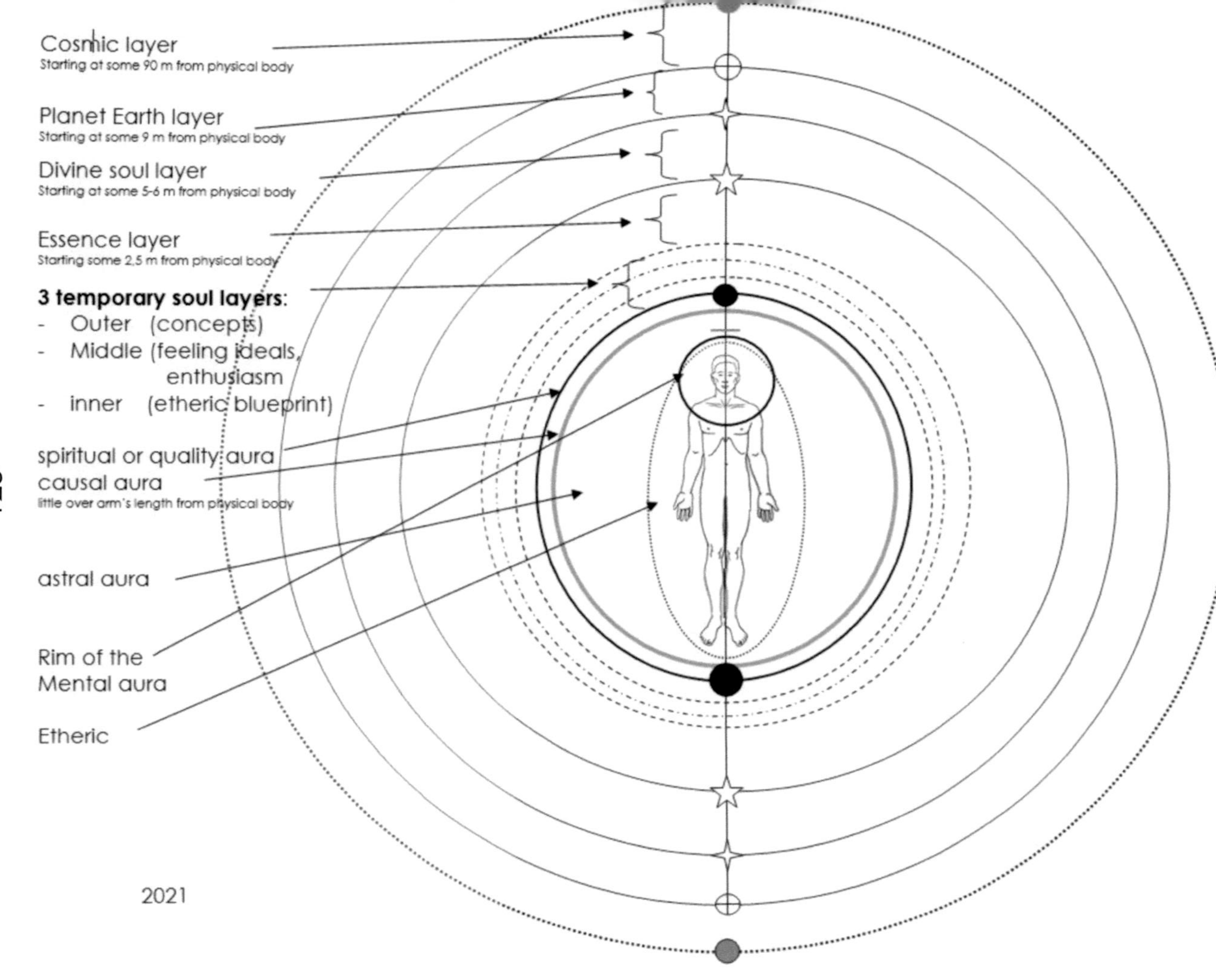
Cosmic layer
Starting at some 90 m from physical body
Planet Earth layer
Starting at some 9 m from physical body
Divine soul layer
Starting at some 5-6 m from physical body
Essence layer
Starting some 2,5 m from physical body
3 temporary soul layers:
- Outer (concepts)
- Middle (feeling ideals, enthusiasm
- inner (etheric blueprint)
spiritual or quality aura
causal aura
little over arm's length from physical body
astral aura
Rim of the
Mental aura
Etheric
2021

Quality colours in the spiritual Aura

Each person will have one, two or (rarely) three quality colours that they incarnate with as their inborn faculties. They are seen and felt in the quality or spiritual aura, which is found at a little over arm's length away from the physical body, in our aura. No two persons with the same colour would express this the same way in their lives. Knowing one's colour, and being able to express it in life nourishes the soul and that colour and quality within. Our 'personal' spiritual energy can be contacted above our heads, a little over arm's length away at the individuality point.

Red	courage, will, power, self-dependent, leadership
Orange	balance, harmony, rhythm, beauty
Yellow	tolerance, patience, logic, precision
Green	impartial, adaptable, instinctive, mental strength
Blue	intuition, love, wisdom, perception
Rose pink	devotion, loyalty, service to others, directness
Violet	truth, ritualism, activity, integration, vitality, dignity

In what is called 'spiritual healing', the starting point or essential reference point is the spiritual level of a person, a community, a landscape. The spiritual or quality aura of a person shows the higher energy that is most beneficial to include in any transformation process.

The origin of impulses

Writing about the intelligence behind nature, brought me to introduce you to a number of many facetted intelligences. This is an unusual attempt as it may bring insights into (quantum?) fields that we know little about. It is a work in progress and I aim to be able to gradually bring out more details and hopefully more clarity.

The following pages are partly from my book 'How to create Divine Art'. [28]

Higher feeling, Divine Field or spiritual plane   (21 levels)
These levels of higher feelings can only be perceived through the upper layer of our reflector ether. As we progress in our personal development, we naturally open ourselves to this perception through the reflector ether.

The spiritual plane is a high source of inspiration coming from beings of the Divine Field. A piece of music for example would also reflect back its influence into the spiritual plane. The following list was elaborated in cooperation with 'C'.

It began when I asked C how many levels were in the Divine Field: 21 they replied and added they would help me to describe them. I started with the nine categories of angels of Dionysius Areopagita * and immediately added level 21 for the Universal Intelligence. We added the numerous angels I had met in nature: landscape and regional angels, nation angels and the angel of planet Earth. Then we added the angels and very big elementals of river systems, deserts, rain forests, etc.

*) Dionysius was the first bishop of Athens and a disciple of Paulus.

Level 17 sees the introduction by 'C' of the ‚Cupids'. These are not found in known lists about the hierarchy of angels. 'C' obviously wanted to include them as they are specifically the high angels of art. In the roman mythology Cupido is the son of Venus who can transform people with his arrows of love and beauty. Obviously, we are referring to Cupids in a wider sense as they are a high level of angels who are specialized in the spreading of beauty, art and love.

The following list of angels does not correspond to what you may find on internet. In his 2nd and 6th century incarnations, Dionysius Areopagita* established his list of nine types of angels and their hierarchies, but mentioned that this list was only an approximate model and that the only one who would know the exact facts was the Creator himself: Seraphim, Cherubim, Thrones, Dominions, Virtues, Powers, Archangels, Principalities, and Angels. 'C' insisted that the 'Dominions' were not the same as the Kyriotetes (Greek word). The Dominions having their own list of tasks.

We need to bear in mind: No matter how much we try to understand the divine dimension, we will only succeed up to a certain degree. Besides the brief description of each sphere, we can partially grasp its field of action by excluding from it what other spheres are doing. Again: the numbering is NOT meant to be a value judgement; it could be replaced by alphabetic letters, colours or symbols.

The Divine Field

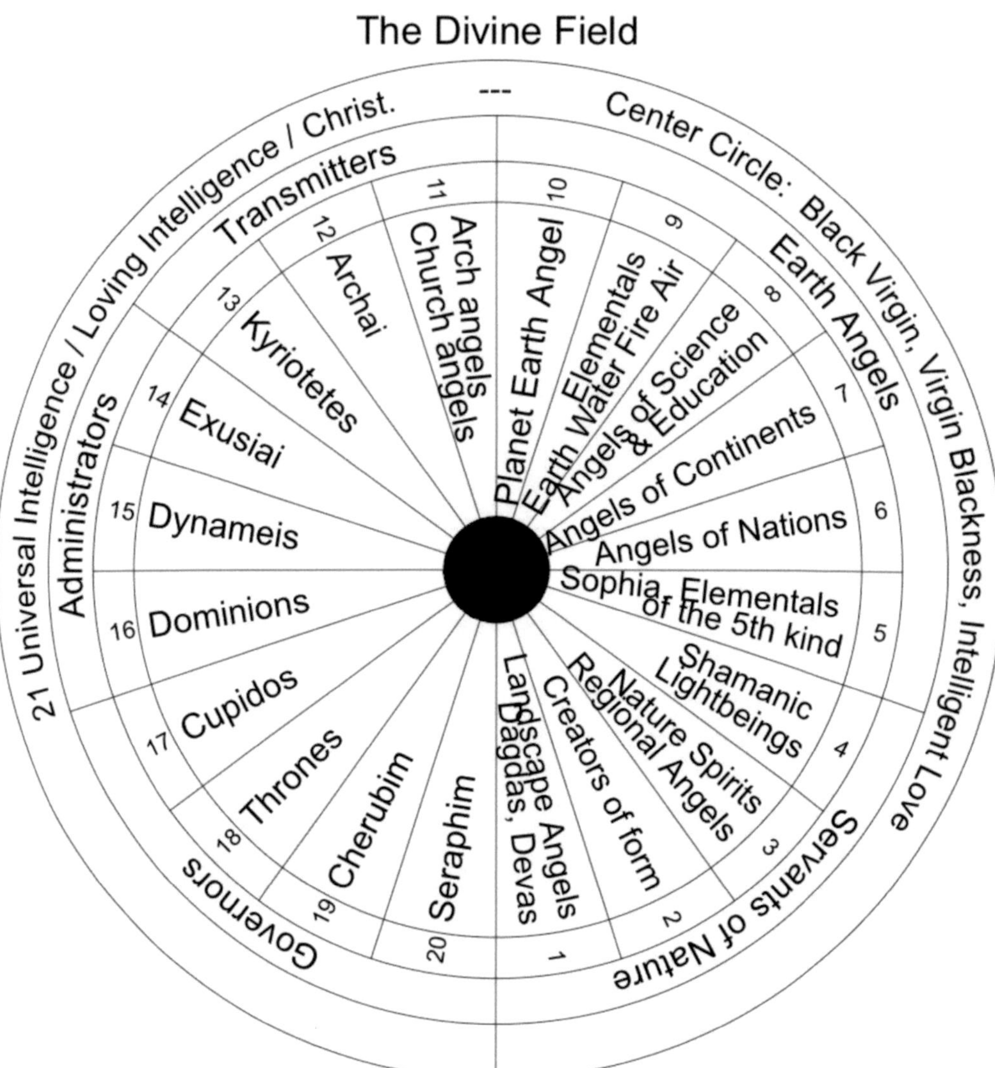

Angels are beings of the Divine Field. I approach them as if I would be looking at a craftsman or craftswoman working with all their skills and wisdom in a completely unknown field to me. This Divine Field includes a number of (quantum) fields and worlds. To begin looking into these fields we need to let go of all we associate with angels, including what the religions have told us, as basically we know very little about them. They are of course an invaluable part of the intelligence behind Creation. They are the inspirations or impulses coming from above, the Heavens, the Father impulse or Universal Loving Intelligence. It is a fact, that it is difficult to get precise understanding of any category listed here. When we look, for instance at sphere 13 beings, called with their Greek name 'Kyriotetes', we can grasp some of their unique tasks by excluding all the tasks listed with the beings from the other 20 spheres. Kyriotetes for instance are not primarily about bringing the impulse of Beauty and high Arts (of sphere 17), nor about Divine Will, as described under No 18, etc.). Looking at the Kyriotetes this way, by eliminating all the other attributes, gives us a clearer idea what their task or impulse is.

The wheel diagram suggests that they are not part of the Administrators, those beings who organize the Divine Field and bring the impulses down to the next level, the Transmitters, beings that bring the impulses further down to humans and all beings below the sphere 13.

What I call the 'Governors' are often referred to as 'the upper or first Hierarchy, the 'Administrators' as the 'middle Hierarchy' and the 'Transmitters' as the 'lower or third Hierarchy'.

1   - Landscape or area angels
     - Dagdas & Devas
2   - Creators of form
     - angels of small lakes, small rivers
     - smaller deserts (a.o. Namibian desert angel)
3   - Wisdom and experience of nature spirits
     - Kali (breaking down & renewal)
     - angels of larger lakes and rivers systems
     - regional angels
     - angels of larger deserts (Gobi & Sahara, etc.)
     - the 72 angels
4   - Inspirations from shamanic light beings
     - Indonesian tropical forest angel
     - guardians of sacred sites
5   - Sophia (wisdom)
     - wisdom of animals, the tall Elves
     - angels of large tropical forests (Amazonian, equatorial African)
     - elementals of the 5th kind
     - Ocean beings incl. earth, water, wind elementals
6   - Nation angels: help shape the spirit of a nation
7   - Angel or spirit of a continent: Europe, Middle East, North America, Arica, etc.
     - Dakinis; Fairies translating humans/nature spirits
8   - Angel of science, knowledge & education
     - underlying structures of life
     - ocean angels: Atlantic, Pacific, etc.,
     - smaller mountain range: Pyrenees, Appalachians
9   - St. Bridget - Goddess of Spring and Poetry
     - the major Elementals (earth, water, fire, air), volcanoes
     - larger mountain range angels: Ural, Himalaya, Alps, Andes, Rocky Mountain, etc.
10 - Planet Earth angel

11 - Church angels
- Archangels: Prayer/Sound (Sandalphon); good
  news & creativity (Gabriel); hope & aspiration
  (Ramiel); service (Jehudiel); faith & spiritual
  strength (Saraquiel); justice & harmony (Raguel);
  wisdom (Raziel); transformation of the shadow
  (Binael); of healing (Raphael); teaching (Uriel),
  Light into shadow (Michael); Divine will (Hesediel)
12 - Archai – Lordships, Principalities : leading the earth
  leaders, people and communities
13 - Kyriotetes : instructing the duties of the angels
  below them; their energy is pure grace;
  - teachings of aliveness, vitality, and joy of life
  - monitoring and participating in earth healing e.g. for
  very large Elementals
14 - Exusiai - Powers : protecting the heavenly spheres from
  all negative influences of the earthly spheres. Keep the
  world in balance and especially the balance with the
  dark forces. Directing the regional Angels
15 - Dynameis - Virtues : directing the cycles of the
  planets; celestial harmony;
  architects of the light ether structures
16 - Dominions : leading the angels of the earth,
  of continents and nations
17 - Cupids : high angels of the Arts, Love and Beauty
18 - Thrones : angels or spirits of divine will and life
  energy, giving impulses, structures and direction for
  humanity
19 - Cherubim : angels or spirits of harmony and wisdom
20 - Seraphim : angels or spirits of light & fire, igniters
21 - Black Madonna*, Christ, Creator, Holy Spirit,
  Buddha, Universal Intelligence

*) I have expanded on devas, landscape and area angels, angels of nations, the whole spectrum of non-physical beings, the human aura, layers of the soul, in "Faith is the Bridge" [29].
www.engelwelt.de and other webpages on this theme of angel hierarchies

Mental plane   (32 levels)   1-9 is ego / 10-20 is 'intuition' / 21-32 is inspiration from the divine field

The mental plane can be understood as a separate world existing in another dimension. It can be a source of mental inspiration through concepts, ideas and visions (e.g. sacred geometry). Art can also reflect back and bring influence into these mental levels. The impulses come from spiritual guides and masters who all have been incarnated on earth before.

As we can see in the list below there are fundamentally different levels within the mental plane. Level 13-32 are part of the 'upper mental', level 1-9 are representing the ego level or 'lower mental' [28]. This type of mental activity is fuelled and often heavily tinted by painful emotions; that is the infiltration of emotions into the lower mental aura. I will not be expanding on this here any further [3].

13-16 ET's information
    13 – ET technology
    14 – ET energy knowledge
    15 – ET wisdom (philosophy, social matters, governance)
    16 – ET visions for the future

17-20 light beings and angel information
    17 – angel of a nation
    18 – energy alignment and balance
    19 – thought structures coming from the spirit of time

20 – angel of groups of nations (Europe)
    conceptual level of creation

21-24 inspirational thoughts from archangels on
    21 – teachings of personal transformation, bringing light
        into darkness
    22 – education, spiritual training
    23 – healing, faith, divine will
    24 – devotion, service, compassion

25 – spiritual issues taught by principalities
    (see p. 135 level 12 of angels)

26 – highly evolved spirits like 'C', masters, etc.

27-32 divine inspiration
    27 - nature of light
    28 - Underlying concepts of harmony
    29 - Underlying concepts of beauty
    30 - nature of non-duality
    31 - nature of love
    32 - spirit of evolution, progress and innovation

It can be a lifetime's work to access levels beyond the ego. I have described many aspects of that transformational work in a number of books. [3]

Astral plane   (24 levels)

Impulses from the upper astral come from astral beings. They live in a totally different world or quantum field, called the astral world or astral plane. These astral beings have never been incarnated on earth, according to C. They are not from the angel hierarchies neither.

1-12 lower astral (painful emotions) / **13-24 upper astral** (feelings)

 1 - fear
 2 - dishonesty
 3 - lust
 4 - anger
 5 - hate
 6 - feelings of inferiority/superiority
 7 - greed
 8 - jealousy, envy, competitiveness
 9 - false pride
10 - sarcasm, cynicism
11 - self pity
12 - sluggishness, self-indulgence, lack of clarity of
      Motivation

----

13 - success, acceptance of present circumstances &
      bases of a situation, 'easy is right' or a naturalness
14 - surrender to the universal or divine intelligence
15 - calmness, serenity, naturalness, humbleness
16 - love, understanding, truth
17 - joy, celebration of life and Creation
18 - compassion
19 - faith, trust
20 - ecstasy, delight
21 - unity, softness
22 - freedom of mind and beliefs
23 - knowing, beingness
24 - eternity, harmony, love

The human etheric   continued

Bob Moore taught extensively about the transformation process in human beings. His comments on the etheric are therefore centred on the human etheric. Many characteristics of the etheric are of a general nature and are shared with other aspects of life. We are part of nature. In the following pages I will share quotations from Bob Moore about the human etheric which can help us to experience and understand its different aspects.

I cannot say how the two etheric layers inside the body are exactly positioned. Bob explained that the four layers were interlaced. He said that certain points, like the secondary chakras, the 1st depression point **dp1** or the four polarity points (shoulder points **sp**, hip bone points **hbp**, see drawing page 142) had a connection to all four layers, whereas some points in the lower part of the torso mainly had a link to the chemical ether, due to the fact that they are strongly linked to elimination (digestion, menstruation). **3**   see also p. 107 for light ether positions

The etheric stores any impression we get from our senses, without discrimination. It does not have an intelligence like the astral and therefore cannot select what it wants to store or not. Access to the stored memory in the etheric, as well as the control of the etheric can for humans only operates through the mental level. Bob Moore: "The control of etheric storage is only possible through the mind level. The area of control are the 4 upper chakras but centred in the pineal. This control can be interrupted by the navel area." The navel area is a place where we can hold many emotions, mainly fear.

The etheric plays an important role in the integration of the different energy layers and levels of consciousness. It governs the overall circulation of energy from head to toes and brings pranic energy from outside the body into the cells of our body with the help of the chemical ether. On a consciousness level this process corresponds to allowing external experiences to penetrate 'under our skin' and then to become aware of how they feel deep inside. In any natural state this would trigger a true reaction on our part which would then be naturally expressed towards the exterior.

Bob: "We often keep an experience outside. When we progress, we let the experience get inside us."

This natural process is interrupted, when on one hand we don't let external experiences come in, and on the other hand when we don't express what we feel about those experiences. This then creates a split between the etheric inside and the etheric outside the body, between the truth & light ether and the life ether. We can look on this as a lack of truth and if it is lasting, this split creates an energy blockage. Overuse of alcohol and drugs has the same effect.

We can detect such zones in the etheric when we slowly draw a felt line (using the mind *and* our feeling perception) on the skin surface and can then meet those places. Directing mental and feeling energy towards these areas gives us the means to connect to our feelings about an experience that up until then, we may not have been able to process. Once we reconnect to our feelings, our consciousness then heals the split in our etheric. We become realigned with the truth of our own feelings.

This **life ether** is linked to the element of earth and everyday life. It extends about12 cm out from the skin surface. It feels denser than the reflector ether. I sometimes call it the 'daily life ether' as it co-ordinates and remembers the activities of daily life in as much as it allows us to do certain routine activities almost automatically, without having to think about what we are doing. We can see this happening when we drive several hundred meters, on a road we know very well, and think about something completely different.

This is possible because the life ether has stored all the information for that specific routine activity many times. The following quotes between "marks" are from Bob Moore.

"The ('daily') life ether reflects the polarity situation between male and female, positive and negative. It is strongly present in the zones of procreation and involved with menstruation and its potential for purification. You find it also predominantly around the female breasts. The life ether plays an important role in our physical health and withdraws gradually when a person reaches the death process. It is connected to how we accept ourselves. Its symbol is the square."

The human etheric

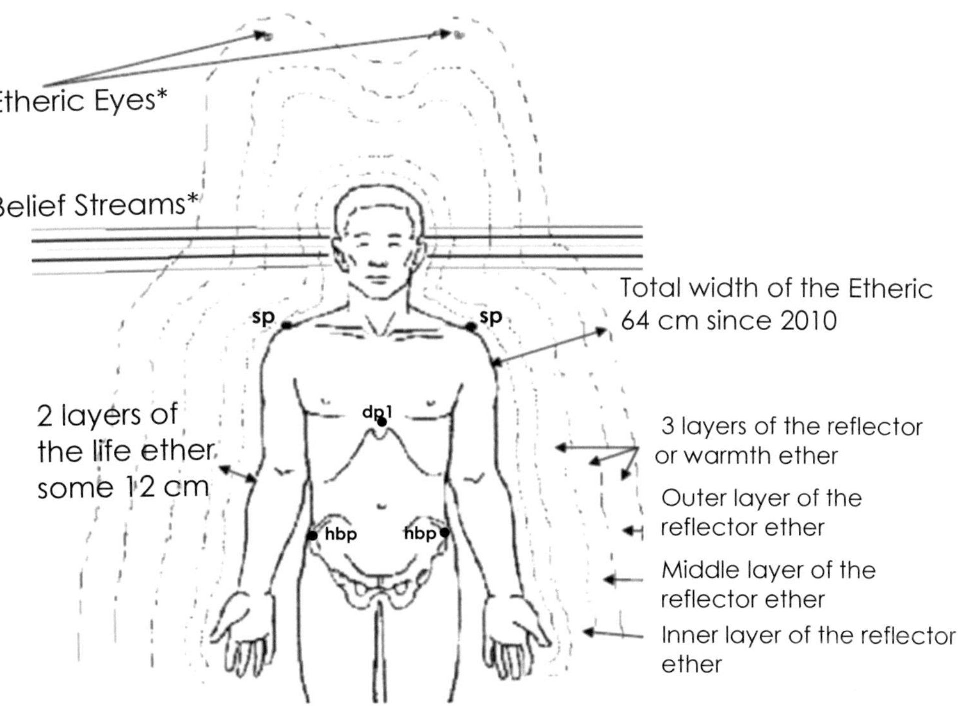

"The reflector (ether) attracts what the chemical ether needs. This **reflector ether** also has the function to reflect towards the exterior what we feel deep inside us. Thoughts from the exterior (e.g. intuitions) penetrate our system here before entering the brain. This layer gives us the means to see colours, to see the aura, to feel the quality of contact between people. It contains also the world memory*, wisdom, knowledge, the physical memory."

*) For more explanations see my book 'The Science of Spiritual Healing'[3]

It reflects what higher energies are bringing into the physical and into our thoughts (intuitions, in the heart- and chest-area).

"Everything that is taking place can be reflected outward. You can use the outer part of the etheric as a means of reflecting out emotion (anger or suchlike) or creating your own projections. That is being done from what you would be retaining within yourself, related to your emotions. There is not necessarily a feed in - to reach that - from a higher level. Maybe you are just projecting out again your own reflections back from other people's emotions. That may mean you are separating the outer layer of the etheric from other layers. So, you are just using the outer layer as a reflector backward from what other people are saying to you. You don't want to be hurt so you just throw it back to them again. You are creating a sort of a vicious circle. But that of course is not helpful to the etheric itself nor to the other levels of the etheric which need to be incorporated if you are going to be using say such things as compassion, which is also a reflector."

We can understand that this process of reflecting is vital and is always linked to our expression. The root chakra and the thyroid chakra, being the two main chakras involved in expression, they are strongly linked to the reflector ether and have an important part to play in its movement.

"The physical level has integrated in it the etheric. The etheric is a storehouse of a large amount of activity which is often reflected from the etheric into the physical, either as blockage or as movement of energy."

"The centre of the head has the controlling aspect of the etheric movement on each side of the body, the etheric circulations. All are controlled from the centre of the head."

"Everyone has energy movements. If we can learn how to utilise the chakras better, if we can learn how to work with the energy that is in the etheric, then we will be better people because all that is going to bring you to a deeper contact with yourself related to what you believe in."

"Chakras are situated on the etheric structure and the etheric structure preceded the physical structure. Chakras therefore were placed into position before the physical structure was completed.

The etheric is in reality the **subconscious mind** and is **a store house**, so we can find that in the etheric we have the memory of past experiences which are already stored there.

When you relax, the etheric expands but you have to be relaxed to allow the expansion."

As I mentioned before: when a person knowingly tells a lie, it will always affect the etheric and create a division between the etheric inside the body and the etheric outside the body and the whole energy frequency will drop.

"When the vibration drops the balance of the **chemical ether** that is needed to support the inner part of the etheric, it is no longer accessible. It's not possible to draw it in because the levels of the etheric are separated. The lower part of the etheric inside of the physical body, that we could call the more physical part of the etheric, needs the chemical ether structure to support it and to support the connection to the physical body. If it cannot receive that then of course you get tired, depressed, something is going to happen physically with you in the way of illness, maybe not serious but something will happen."

"In this way then, the movement of the 4 levels of the etheric are needed to be in a constant frequency of activity to allow a person to remain healthy in their own thinking structure as well as in relation to what is taking place in the physical body, in the frequency of the vibration that is showing up around them.

If you look at yourself and you find there are some **hereditary factors** that you can't change, you can always learn to accept them and in learning to accept them you can still draw on the chemical ether structure that surrounds you."

There are many aspects of the human etheric not covered here: the etheric eyes, the 8 streams and the overall energy circulations on each side of the body. [3]

"Our bodies are made up of 60% water and in that we have also electricity. Each of us has this electricity which surrounds us, so the electricity and the water coming together creates what they call conduction. That moves into our etheric and becomes vapour, because of the heat of the body and that has its effect on chakras. It's the water content with the electricity in it that allows the chakras to rotate because the chakras rotate to the polarity of the earth.

The consciousness of memory is held in the etheric, remembering that the etheric is really the subconscious level of mind, can bring forward things which can really be tremendously helpful and important in relationship to recognising your own process in your development."

"When we look at ethers that are reflected into our etheric structure, then we have the four levels of ether, just as we have the four levels of the etheric structure. These four levels are so divided that they make up a harmonizing situation that is found to operate in the looseness of the etheric, which shows up in the expansion of the mental aura."

"With the **light ether** we have the symbol of the triangle which connects to balance and light. Light of course is something that is necessary and essential to have drawn into a state where there is, for example, depression. A person is not going to release depression until there is light being drawn in, and they have no means of drawing light in themselves, and this of course is where healing is so valuable, that the person who is the mover of energy has the means, because they themselves have inbuilt this structure in their process, in allowing that to be passed over to another person. This triangle is tremendously important in its reflection of light onto another person who would be in this depressed state. I am using depression as an example; other states can respond to that as well. And then we come to the circle, which of course is a continuity, everything brought into its circulation, everything brought into its focus, and here is where we have reflection or warmth. This circle is reflecting all the things that we need in transferring from one person to another person, and that always has to begin in the heart."

"The **chemical ether** is very much related to the transference of colour that is reflected to the physical body and this is why the use of colour can be so helpful to make changes in the body; but then we have got to remember that other ethers can also play their part, like **light** itself, and so, when we use this term light, of course this means something different from person to person, because none of us will understand light in the same way, and so, when we reach a level of consciousness that is allowing us to relate more to **higher consciousness** then our appreciation of light changes, and so that change in our appreciation of light then, allows the light reflection or the light movement to become more balanced or useful. The etheric itself responds and so then we have a healthier etheric or a better distribution of prana energy in the etheric."

…the **health aura** being seen closer to the body between etheric and physical…
(DP: At the moment I understand the health aura as being a reflection outward of the state of health of the organs, comparable to heat emanating from an organ. It is not an etheric layer.)

The **life ether** has got two sublayers. The lower layer, next to the skin, is where **our physical senses** are. Their perceptions go into the light ether to nourish the chemical ether and the cells, the **memories** of our perceptions are all stored in the upper life ether layer, for the rest of our lives. This layer is also known as the (here individual) Akasha chronicles. 'Akasha' being the Sanskrit word for ether. Our **higher senses** (clairvoyance, etc.), are located and perceived in the upper reflector ether, while our spiritual or **higher feelings** are senses of the soul level, coming in through the upper reflector ether.

Feelings operate on different levels: from near physical body sensations, to instinctive basic emotions, to refined feelings and up to subtle, spiritual, or higher feelings.

The chemical and the light ether are occupying the same space inside the physical body. They do not function like the layers of a cake but are similar to two **quantum fields** with different functions and laws. While the light ether is everywhere inside the physical body, the chemical ether is also everywhere in the process of creating cells; it is more particularly found around certain areas linked to assimilation and elimination.

It seems to me that each of the four ethers could belong to a different quantum field and thus operate under different laws and properties. That is why the elementals operating in the chemical ether cannot communicate with the elementals of the life ether.

The potential of the reflector ether

Let me repeat this quotation from Bob Moore because of its far reaching implications:

"The reflector ether is the highest ether, the highest energy contact that you can achieve. In that energy contact you have got all the things that are necessary, you have peace, contentment in yourself, you have your connection with the universe, you have a relationship to light, you have what is drawn from that light into the body through the chemical ether and then you have the balance of life which you will find in the balance with nature and yourself. The reflector ether is reflecting all of that as a combination, but centred really on the peace."

I gradually realized that this passage contains great wisdom, pointing to the fact that we need time to progress towards these connections:

'…the balance of life …balance with nature and yourself.'
'…your connection with the universe.' Which lets us understand the reflector ether in our aura as being the interface with the Divine Field.
'…but centred really on the peace.' The peace in ourselves and as part of All-there-is.

"All this energy that we are talking about works the same with everyone, but it is never brought into a focus with many people. The reason that it is not brought into focus is because the ether which is the highest ether that is known about, which is called the reflector ether, is not in evidence, because the person is too occupied with things that they are doing with their own intent, that is with their own physical will. Once a person brings their own physical will into doing things, especially with healing, then of course there cannot be this opening, and so all this energy movement is there and this is what can be quite deceiving. The energy movement can be there but of course if it is not created in its movement of expression with the reflection or the reflector, then of course it cannot be used to purpose and so, in this increase of movement of energy, that increase cannot come about by physical will. It can only come about when the awareness has been allowed to reach this higher definition of ether substance and when that again is drawn downward, then it increases the potential and that is the quickening up of energy. That in reality is what is talked about in the bible as the BREATH OF LIFE, that is what it is."

I discovered that the effects of the sound clouds of harps and harp like instruments (hammered dulcimer, sitar, zither, etc.) travel in this upper layer of the reflector ether…around the whole world and their effects reaches the reflector ether of individuals receiving distant healing.

Transformation of frequencies in the human aura

I do not know exactly how the transformation from one type of energy into another happens. Yet, we can observe a general rule, that higher frequency energies penetrate and, to a certain degree, transform lower frequencies. As a general statement we may observe that the energy of Love or loving kindness is the fastest energy. It is said to be the strongest energy available and that it can penetrate everything. The experience of working on oneself certainly shows this to be true.

We can experience how the combination of patience and loving attention, which is consciousness, does affect denser energy fields, like painful emotions or body tensions and pains. This is also the case with our etheric, although becoming aware of this phenomenon may require some practice of observation. We may have the experience how spiritual thought and feelings, coming clearly from beyond the ego level, can change our beliefs, thought patterns, individually and collectively. This type of thought may for instance uplift the energy in a group of people. Whereas the constant dwelling in problems may not help to find solutions.

We all can observe how our painful emotions can at times influence our thoughts. We can ask why this is, as emotions are of a slower vibration than mental energy. A healer colleague told me that when working with a client, she had perceived on several occasions, that their causal aura suddenly opened and showed the belief structures, which are thoughts, that allowed their emotions to enter the lower mental. Thus, although we can see emotions penetrating the lower mental, it is actually thought patterns in the background (causal aura and beliefs) that allow this to happen.

A little side walk for those interested in sound healing
As a curious sound healer and harpist, I wondered how the actual transformation through frequency would happen. Sound seems to be the vehicle, that transforms divine energy into the physical, while colour is the nourishing part of that energy. This happens essentially by means of the resonance phenomenon of the musical 5th.

On page 116 I described flying insects that were producing a buzzing sound around a tree all at the same frequency of 256 Hz. This sound contributes to convert the reflector ether energy of 768 Hz into life ether energy of 256 Hz. 256 is the note C (when using natural tuning of 432 Hz for A). Its strongest harmonic or overtone, after the octave, is its 5th which is at 768 Hz (256x3). It is found by playing a bell tone or harmonic at 1/3 of the C string. This corresponds to a G 1½ octaves above our C. The information I get is that 768 Hz is the key frequency or central address of the reflector ether. 2304 Hz being the central address of the astral Aura, etc. Making a G string ring on a harp will by this resonance phenomenon activate the C 1½ octaves below. Through that a higher note or frequency transmits energy to a lower frequency. This natural resonance creates the 'wire' or tool that can bring down the celestial reflector energy. This phenomenon explains why harps (or any multistring

instrument) are particularly designed to bring down celestial energies. It also brings a new light to the 'undertone or aulos scale'. [22]
I will now introduce my source of information 'C', the college of spirits of Rocamadour.

My source 'C' - The **C**ollege of spirits of Rocamadour. Rocamadour is an ancient pilgrimage site less than an hour away from my home. I came in contact with 'C' ten years ago, and they participated in eight of my books. My communication with them is pragmatic and based on YES/NO questions. I always received immediate and precise answers.

In my communication with invisible beings, I have worked out a rigorous check list of requirements that need to be fulfilled, a check list that actually I also use for communicating with humans:

Any communication needs to have coherence, consistence, new and unexpected aspects, useful, prove to be of good common sense, respecting my free will and to be precise.

I wanted to know the composition of the spirits of Rocamadour as I needed to find out who I was in contact with. 'C' consists of 12 honorable permanent members and a wider circle of experts. Many of them are scientists who can draw upon specific knowledge when a subject requires it. I always get a precise answer about anything I have asked them.

I presented the 12 permanent members in a previous book. I asked them whether they wished to be named in this book and they answered that naming them helps to bring clarity and demystifies who they are.

9 of them have previously been incarnated on earth, 7 as Christians: Pope Leon IX (11th century), 4 Saints: St. Amadour (1st century), St. Alain de Lavour (7th century), 2 female Christiane de la Sainte Croix (IT 14th century), Sainte Hildegard of Bingen (12th century), a Tibetan Buddhist Rinpoche Lama Guendune (he was leading a monastery in Dordogne until 1997), 1 scientist: Marie Curie (Nobel price), 3 were never incarnated on earth: the Deva Queen of the SW of France, 2 from 'outer space'*: a teacher from Pegasus, 1 female teacher from Arcturus. *) We actually don't know what that means.

Until the year 2000 C their main task was to teach the souls of deceased people who had returned to Rocamadour, a holy place that they had known during their lives. Around 2006 C started to expand their field of action considerably. This led to the present day organization as a Think Tank, a university of ideas and knowledge. C is not attached to any particular religion.

I am including the drawing next page, not that I think you could fully understand it right away – it took me more than 30 years to get there – but because I want you to see how precise the science of subtle energy can be and that it will lead us, together with quantum field theory, into a broader understanding of life.

The Etheric layers outside the body

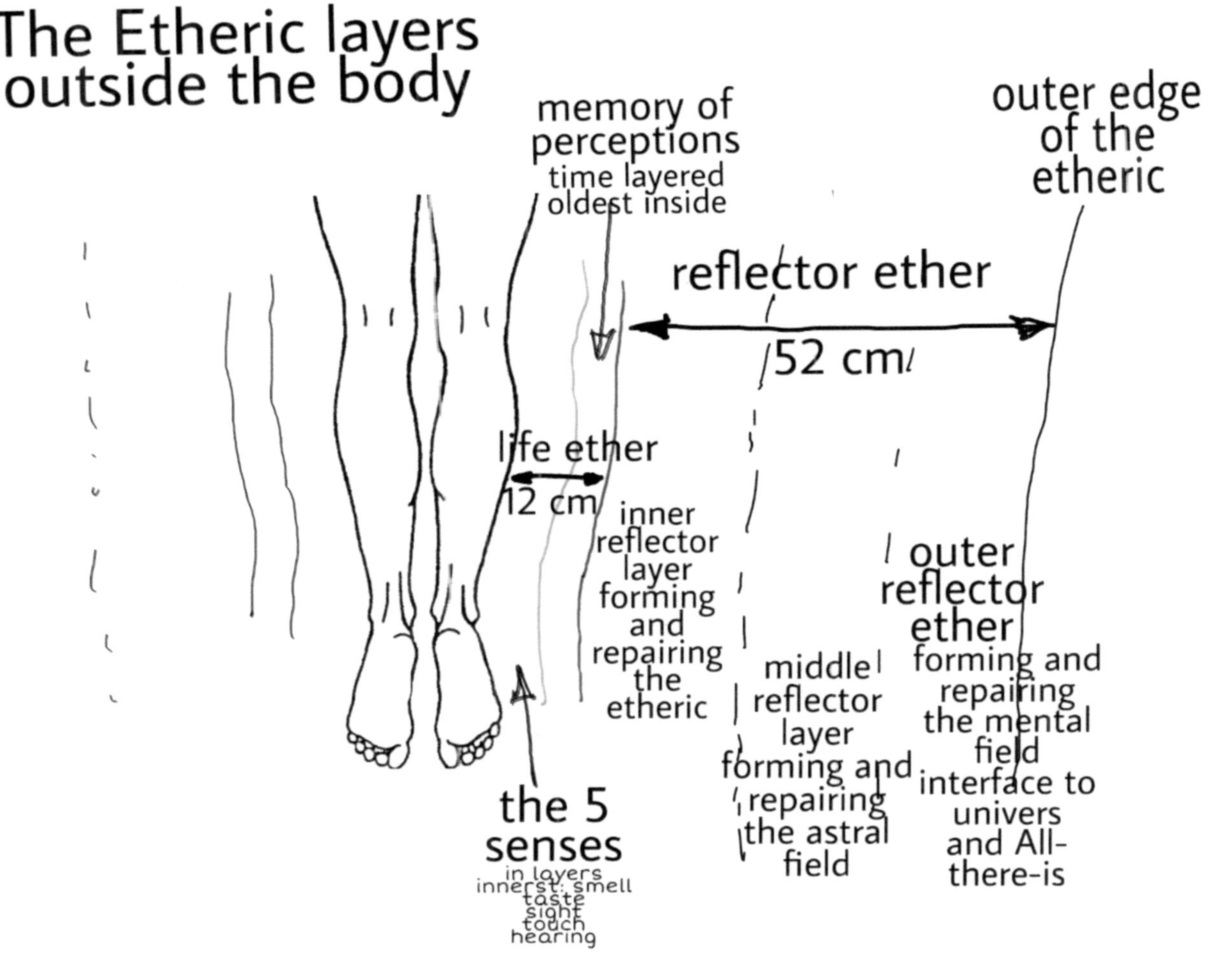

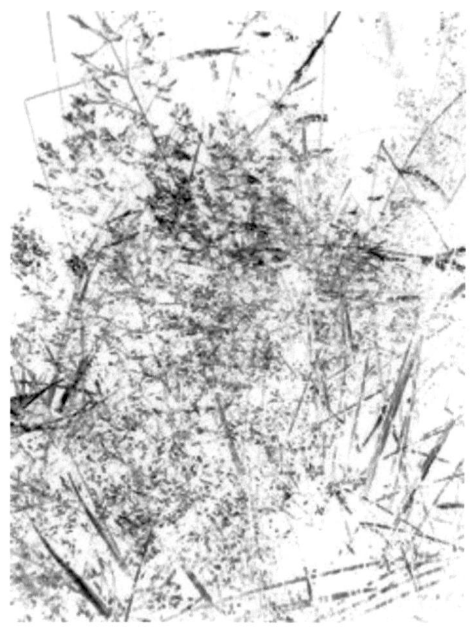

Interview with the Mother Earth Being

Is it ok to call you 'Mother Earth Being'? Yes.

The intelligence behind nature is a complex phenomenon. It is impossible to do it justice here in the framework of a book. Yet, it is necessary to become as precise and as concrete as possible, so that we can appreciate and respect who we are communicating with. (The being is ok with this statement)

Can a dialogue with larger nature spirits in charge of areas contribute to any political decision process about the area? Yes.

Can this dialogue be expected to be like an interview with a human being? No.

Would the respective nature spirits in charge of an area or part of an area be expected to bring forth a dialogue partner representing them all? Yes.

When working towards the recognition of the rights of a whole river system or a wet land area, could the group of people working on the question count on finding a competent and representative partner amongst nature spirits? Yes.

Would that nature spirit be expected to make politically understandable contributions? No.

Do you mean 'not necessary'? Yes.

Does that mean the nature representative would come up with its own range of requirements and information? Yes.

So, their contribution could often be unexpected? Yes.

Would it always be a valuable and essential contribution to the political decision process? Yes.

When I use the term 'political' process can this include also possible legal procedures? Yes.

Can we expect that such a dialogue would attract other conscious beings that would help uplift the energy level of the process? Yes.

This rise in energy level would include more tolerance, understanding, compassion, patience, joy, respect? Yes.

This would help to bring about more mutual understanding? Yes.

It would obviously bring in unexpected aspects of the natural area, helping to appreciate more its complexity? Yes.

Would it also help in raising more understanding of how the intelligence behind nature functions? Yes.

When not including a dialogue with the intelligence behind nature in such a process, would that mean often having more difficult procedures? Yes.

More misunderstanding? Yes.

More ego battles involved? Yes.

Including nature spirits in the process would help diminish these issues? Yes.

Can one say, that including nature spirits in the process would automatically bring in 'spiritual energies', the good will and experience of higher being and thus raise the whole level of consciousness and cooperation? Yes.

So, although the process might often bring in some unexpected angles and statements, the overall result would be significantly positive? Yes.

This means that the humans involved would usually need a greater tolerance towards an unusual reasoning? Yes.

Do you feel we should involve other aspects in this present interview? No.

Are you pleased with our dialogue? Yes.

Do you feel I have understood you point of view correctly? Yes.

Do you feel you have participated already in the formulation of the questions? Yes.

Thank you for that dialogue. I must say, I had not expected you to answer.

Using a Hartmann-Antenna

If in quantum physics they can accept the phenomenon of quantum entanglement*, a force we cannot completely understand yet, we might be able to accept that there are comparable forces in radiesthesia that move a tool like the Hartmann-Antenna to show YES and NO answers.

The communication between human beings and nature spirits or beings of the divine field operates with the help of a translator being. They are sphere 7 beings. For making a communication with invisible beings possible, human beings need to connect with their feelings, then produce their question. This question is being translated by these sphere 7 beings. The invisible beings then respond with their type of feelings which then are again translated. The interface in humans is our upper reflector ether.

*) **Quantum entanglement** is a physical phenomenon that occurs when a group of particles are generated, interact, or share spatial proximity in such a way that the quantum state of each particle of the group cannot be described independently of the state of the others, including when the particles are separated by a large distance. The topic of quantum entanglement is at the heart of the disparity between classical and quantum physics: entanglement is a primary feature of quantum mechanics lacking in classical mechanics. (Wikipedia)

By way of concluding, I wish to share that a large elemental of the 5th kind, specialized in the cooperation between nature spirits and humans, contacted me and asked me to add this:

'It is within reach of every human being to be able to communicate with nature spirits. Humans can simply begin by asking them a question about some aspect of nature. Asking a question is the main step, as it shows that one respects the other and shows interest in being open to take into consideration their existence, intelligence, and opinion. Any person can, with their very own means, get an answer: with their feelings, their intuition, a vision, a radiesthesia tool or a more sophisticated channeling consisting of complete sentences.'

I describe my communications with nature spirits in my book [16]

We *are* nature,
but thought for too long that
we were a separate part of it.

References

1) The Christ Letters, Free download from: www.christsway.co.za
https://thechristletters.weebly.com/uploads/1/2/5/7/125746987/christ_letters
_original_version.pdf – printed book at https://www.christsway.co.za/
2) Wall Street International Magazine WIM, Series from 11 DECEMBER 2020 –
11 April 2021, Federico FAGGIN: The nature of consciousness
3) D. Perret, The Science of Spiritual Healing, BoD, 2010
4a) Verena Staël von Holstein, Wolfgang Weirauch 'Nature Spirits and What
They Say' Interviews, Floris Books, 2004, ISBN 978-0-863154-62-1
4b) 'Nature Spirits of the Trees', Interviews Verena Stael von Holstein, Edited
by W. Weirauch, Floris Books, 2009, ISBN 978-0-863157-03-5
5) www.etimo.it/?term=natura
6) Archiv zur Geschichte der Max-Planck-Gesellschaft, Abt. Va, Rep. 11
Planck, Nr. 1797
7) Prof. Dr. Suzanne Simard, 'Finding the Mother Tree – Discovering the
Wisdom of the Forest', Allen Lane publishers, May 2021
8) Peter Wohlleben, 'The Hidden Life of Trees: What They Feel, How They
Communicate', Greystone Books, 2016
9) Stefano Mancuso, Alessandra Viola, 'Brilliant Green: The Surprising History
and Science of Plant Intelligence' Island Press, 2015
10) Wikipedia free, open-collaborative online encyclopedia
11) Linda Gottfredson, (1998). 'The General Intelligence Factor' (PDF).
Scientific American Presents. 9 (4): 24–29. Archived (PDF) from the original on
7 March 2008. Retrieved 18 March 2008.
12) David Wechsler, (1944). 'The measurement of adult intelligence'.
Baltimore: Williams & Wilkins. ISBN 978-0-19-502296-4. OCLC 219871557.
13) Humphreys, L. G. (1979). 'The construct of general intelligence'.
Intelligence. (2): 105–120. Doi: 10.1016/0160-2896(79)90009-6.
14) Dan McKanan, 'Eco-Alchemy', University of California Press, 2018
15) Susan Borowski in Blog 07/16/2012 of American Association for the
Advancement of Science
16) D. Perret, Erd-Heilen / Guérir la Terre, BoD, 2018
17) D. Perret, Distant Healing with the Harp, BoD, 2020
18) D. Perret, Contributing to Healing, BoD, 2020
19) Dr. Jacques Mabit, 'Amazonian shamanism and the Western world:
between encouragement and warning',
www.takiwasi.com/docs/arti_ing/amazonian_shamanism_western_world.pdf
20) D. Perret, L'accès aux mondes invisible, BoD, 2017

21) Consciousness and its Place in Nature, David J. Chalmers, http://www.consc.net/papers/nature.pdf

22) D. Perret, Music as a mystical Journey, BoD, 2013

23) WHYY podcast July 30th 2021 Kenny Cooper, Zunigha is an enrolled member and cultural director of the Delaware Tribe of Indians, one of three federally recognized tribes of the Lenape in the United States. WHYY is a public media organization in the Philadelphia Region, including Delaware, New Jersey, Pennsylvania

24) https://wcmcoop.org/Wisconsin citizens Media Cooperation; Rights of Nature: Giving Voice to the More-Than-Human, August 10, 2021, Chelsea Fairbank and Barbara With

25) 'Irish Spirit', Ed. Patricia Monaghan, Wolfhound Press, 2001

26) Vanessa Doutreleau, co-auteur of 'Mémoires d'Islande', ateliers des Brisant 2011, scientific advisor organizing exhibitions in the Écomusée de Marquèze), Ethnologie Française, Vol. 33, of 2003/2004

27) C. Lecouteux, Dictionnaire de mythologie Germanique, Imago 2014

28) D. Perret, Creating Divine Art, BoD, 2016

29) D. Perret, Faith is the Bridge, BoD, 2015

The Quotes from Bob Moore are taken from my personal notes during his courses.

30) Grimm, R.L., 'Northern Yellowstone winter range studies. Journal of Wildlife Management 8', 329–334. 1939.

31) Lovaas, A.L. 'People and the Gallatin Elk Herd. Montana Fish and Game Department, Helena, Montana, USA'. Mao, J.S., Boyce, M.S., Smith, D.W., Singer, 1970.

32) NRC, National Research Council, 2002. Ecological Dynamics on Yellowstone's Northern Range. National Academy Press, Washington, DC.

33) Barmore, W.J. 'Ecology of Ungulates and their Winter Range in Northern Yellowstone National Park; Research and Synthesis 1962–1970'. Yellowstone, 2003. Center for Resources, Yellowstone National Park, Mammoth, Wyoming, USA.

34) Ripple, W.J., Beschta, R.L. 'Trophic cascades in Yellowstone: The first 15 years after wolf reintroduction'. Biol. Conserv. (2011), doi:10.1016/j.biocon.2011.11.005

35) www.arizona-dream.com/usa/blog-voyage-usa/la-reintroduction-de-14-loups-a-yellowstone-change-l-ecosysteme/la-reintroduction-de-14-loups-a-yellowstone-change-l-ecosysteme.php

Federico Faggin

Federico Faggin received a Laureate degree in Physics, summa cum laude, from the University of Padua, Italy, in 1965, and moved to Silicon Valley in 1968. He developed the MOS Silicon Gate Technology in 1968; the world's first microprocessor, the Intel 4004 in 1971, and several highly-successful microprocessors, like the Intel 8080 and the Z80 produced by Zilog, his first startup company. Faggin was CEO of several high-tech startup companies he founded and directed since 1974, including Synaptics, the developer of the early touchpads and touchscreens. He is currently president of Federico and Elvia Faggin Foundation, dedicated to the science of consciousness. Faggin received many international awards, including the 2009 National Medal of Technology and Innovation, from President Barack Obama.

I also want to give credit for the groundbreaking work of **Wolfgang Weirauch** of the Flensburger Hefte and Verena von Staël. The 40 books of interviews with nature spirits have laid the foundation of a nature science of the future.

From 1963 to 1978 Christ instructed an English lady, calling herself 'the recorder', in the principles he explains in his '**Letters**'. In 2000, when she was eighty, Christ began to dictate the 'Letters'. The 'recorder' wanted to stay anonymous nor have any financial advantage from the texts.

All photos D. Perret

Photo permission of Yann Arthus Bertrand

danielperret.bandcamp.com

www.vallonperret.com